Richard Watson is an aut⋯⋯⋯⋯⋯⋯⋯⋯⋯⋯⋯ who helps individuals and organi⋯⋯⋯⋯⋯⋯⋯⋯⋯ particular emphasis on strategi⋯⋯⋯⋯⋯⋯⋯⋯ ⋯under of What's Next, a website that do⋯⋯⋯⋯⋯⋯bal trends, and is cofounder of Strategy Insight, a scenario planning consultancy.

His clients have included, among others, IBM, McDonald's, PricewaterhouseCoopers, Virgin, Department of Education, Public Libraries NSW, Ikea, Toyota, Coca-Cola, and PepsiCo.

Richard also writes for a number of business publications worldwide, including *Fast Company* (US), *Future Orientation* (Denmark), and *Retail Banking Review* (Australia).

Richard was born in the UK and divides his time, rather unsuccessfully, between London and Sydney. Apart from two future minds (aged 8 and 10), his other interests include old cars, old wine, and fixing things in sheds.

Further information can be found at:
www.nowandnext.com
www.strategyi.net
www.futuretrendsbook.com

For Georgie, who gives me the space to dream.

Praise for
FUTURE MINDS

"This book is a crusade for slow thinking: a relaxed and funny read that is part guide to the digital age, part compilation of anecdotes and part warning about the seduction of digital speed... A how-to guide to a world where 'an average teen sends up to 2000 text messages a month'.
Watson discusses strategies on how to talk to children, students and young employees, and offers some insights on how to come up with quality ideas.
The book has an informal, more-like-a-conversation feel. It is refreshing to see Watson taking his own message to heart and for the most part writing slowly and thoughtfully."

Australian Financial Review

"The subject at the heart of this book is one that should concern everyone involved in education, namely the effect of digital technology on our thinking and critical skills. Certainly thought-provoking, and the author is incisive in his ability to decipher future needs. The problems are well stated... Controlling technology lies at the heart of the solution and... we need to be aware of the dangers of our current indulgences. We can do little more."

Times Educational Supplement

First published by
Nicholas Brealey Publishing in 2010
Reprinted 2011

3–5 Spafield Street
Clerkenwell, London
EC1R 4QB, UK
Tel: +44 (0)20 7239 0360
Fax: +44 (0)20 7239 0370

www.nicholasbrealey.com

www.nowandnext.com

20 Park Plaza, Suite 1115A
Boston
MA 02116, USA
Tel: (888) BREALEY
Fax: (617) 523 3708

www.futuretrendsbook.com

© Richard Watson 2010
The right of Richard Watson to be identified as the author of this work has been asserted in
accordance with the Copyright, Designs and Patents
Act 1988.

ISBN: 978-1-85788-549-1

Library of Congress Cataloging-in-Publication Data

Watson, Richard, 1961-
  Future minds : how the digital age is changing our minds, why this
matters, and what we can do about it / Richard Watson.
     p. cm.
  Includes bibliographical references and index.
  ISBN 978-1-85788-549-1 (alk. paper)
 1. Cognitive science. 2. Thought and thinking. 3. Information
technology--Psychological aspects. 4. Digital communications--Psychological aspects. 5.
Information technology--Social aspects. 6. Digital communications--Social aspects. I. Title.
  BF311.W294 2010
  303.48'33--dc22

2010033536

British Library Cataloguing in Publication Data
A catalogue record for this book is available from the
British Library.

FSC
Mixed Sources
Product group from well-managed
forests and other controlled sources

Cert no. SGS - COC - 2061
www.fsc.org
© 1996 Forest Stewardship Council

Printed in the UK by Clays St Ives plc.

Contents

"To go faster you need to go slower."
—Sir Jackie Stewart

"The degree of slowness is directly proportional to the intensity of memory; the degree of speed is directly proportional to the intensity of forgetting."
—Milan Kundera

"If we want things to stay as they are, things will have to change."
—Giuseppe Tomasi di Lampedusa

"I don't believe society understands what happens when everything is available, knowable and recorded by everyone all the time. We really have to think about these things as a society."
—Eric Schmidt, CEO of Google

Overture

Screen Culture

This book is partly a sequel and partly a prequel to my previous book, *Future Files*. It's about work, education, time and space, books, baths, sleep, music, and other things that influence our thinking. It's also about how something as physical, finite, and flimsy as a 1.5kg box of proteins and carbohydrates can generate something as infinite and valuable as an idea. Most of all, it's a book of concepts and conversation starters, with 10 key trends as the unifying force, including what I call constant partial stupidity, digital isolation, and a return to the real. It delves into the implications of living in a digital age.

Cellphones, computers, and iPods have become a central feature of everyday life in hundreds of millions of households, offices, and schools around the world. Children as young as 5 spend an average of six hours every day in front of some kind of screen, teenagers and many adults probably more. In the US, for example, adults were spending double the amount of time online in 2009 as they were in 2005. In Europe, the amount of time adults were online grew by around a third over the same time; in the UK, nonworking women are passing almost half their leisure time online, and the average person spends 45 percent of their waking hours on media and communication. Or there's the finding in 2010 that 8–18 year olds in the US spend an average of 11 hours a day in front of a screen, be it a television, a computer, a cellphone, an iPod, or two or more simultaneously.

We are increasingly communicating via text message and email rather than face to face, we have hundreds of online "friends" yet we may not know the people next door, and the first place we look for information is Google.

This technological ubiquity and electronic flood are resulting in significant shifts in both attitudes and behavior, which is what this book sets out to explore. It's about how the digital era is changing our minds—about what's happening now, and what comes next.

But can something as seemingly innocent as a cellphone or a Google search really change the way people think and act? This is a very important question. It is also one that is actively engaging the minds of a number of eminent scientists, particularly those who study the physiology of the brain, because implicit in the question is the thought that the digital age may be changing our brains too.

Michael Merzenich is a pioneering neuroscientist who discovered through experiments that the human brain is "plastic": it responds to any new stimulus or experience. Our thinking is therefore framed by the tools we choose to use. This has always been the case, but we have had millennia to consider the consequences. Arguably, this has now changed, and Merzenich has argued that the internet has the power to lead to fundamental change in our brain, leading it to be "massively remodeled." We are already so connected through digital networks that a culture of rapid response has developed. We are currently so continually available that we have left ourselves no time to think properly about what we are doing. We have now become so obsessed with asking whether something *can* be done that we leave little or no time to consider whether it *should* be done.

For example, according to Professor Susan Greenfield, a brain researcher at the University of Oxford, when kids do something they like, such as playing an electronic game, the brain receives a blast of dopamine in the prefrontal cortex. However, if too much dopamine is produced—if they play too often—the parts of the prefrontal cortex associated with reasoning can be compromised. Electronic euphoria thus creates fewer possibilities and less opportunity to develop an original mind.

The digital era is chipping away at our ability to concentrate too. The quality of our thinking and ultimately of our decisions is suffering. Digital devices are turning us into a society of scatterbrains. If any piece of information can be recalled at the click of a mouse, why bother to learn anything? We are becoming Google-eyed, scrolling through our days without thinking deeply about what we are really doing or where we are really going.

Reading on a computer screen is fast and is suited to foraging for facts. In contrast, reading on paper is reflective and is better suited to trying to understand an overall argument or concept. Both forms of reading—both forms of technology—ought to be able to live alongside each other. Since digital books are becoming instantly available and inexpensive, there is a danger that we will start to view them as just another disposable product, something to be consumed quickly and then thrown away. But if we keep the words and throw away the physical books, we are losing something of great significance, because physical books engage our senses in ways that digital artifacts do not. Reading a physical book is a highly tactile experience that delivers a sense of progression, and printed books give physical and metaphorical weight to the reading experience.

Furthermore, our attention and our relationships are getting atomized. We are connected globally, but our local relationships are becoming wafer thin and ephemeral. We are in danger of developing a society that is globally connected and collaborative, but one that is also impatient, isolated, and detached from reality. A society that has plenty of answers but very few good questions. A society composed of individuals who are unable to think by themselves in the real world.

It is the right kind of thinking—what I call deep thinking— that makes us uniquely human. This is the type of thinking that is associated with new ideas that move the world forward. It is the type of thinking that is inherent in strategic planning, scientific discovery, and artistic invention. It is thinking that is **rigorous, focused, deliberate, considered, independent, original,**

**imaginative, broad, wide, calm, relaxed, attentive, contempla-
tive, and reflective**, where the flow of information is limited and
the medium matters—what you might call "slow flow." It is not
shallow, narrow, hurried, cursory, fractured, or distracted.

But deep thinking like this can't be done in a hurry or an
environment full of interruptions or hyperlinks. It can't be done
in 140 characters. It can't be done when you're in multitasking
mayhem. What happens to the quality of our thinking when we
never truly sit still or completely switch off? Modern life is
indeed changing the quality of our thinking, but perhaps the
clarity to see this only comes with a certain distance or detach-
ment—like when you are sitting quietly to read a book, for
instance.

You might think that none of this really matters; but it does.
The knowledge revolution has replaced human brawn with
human brains as the primary tool of economic production.
Intellectual capital—the product of human minds—is now what
matters most. And we are on the cusp of another revolution, too.
In the future, our minds will compete with smart machines for
employment and even affection. Machines are becoming adept at
matching stored knowledge to patterns of human behavior, so we
are shifting from a world where people are paid to accumulate
and distribute fixed information to a fluid innovation economy,
where people will be rewarded for being conceptual thinkers. Yet
this is the type of thinking that is currently under attack.

So how should we as individuals and organizations be dealing
with our changing way of thinking? How can we harness the
potential of the digital age while minimizing its downsides?
That's what this book is about.

We need to do a little less and think a little more. We need to
slow down—not all the time, but occasionally. We need to stop
confusing movement with progress and get away from the idea
that all communication and decision making have to be done
instantly.

Try as I might, I find it difficult not to be sucked into the vortex of change. To me it feels as if time itself is being compressed. Having even an hour during the day just to think or write, uninterrupted, is becoming a luxury, mainly due to digital technology. I never quite feel as though I am in control, and when I do get a chance to think I usually end up mentally commuting back to an era when things were simpler and more certain.

A study by the University of California (San Diego) has found that in 2008 the average person's daily intake of information was 300 percent greater than in 1960. I don't know about the exact percentage, but I am certainly faced with an avalanche of information every day and I'm on a digital diet of continual deletion. But like any slimming diet it's difficult to stick to for long, so the digital binging goes on and the megabytes keep building up. And while the vast amount of information at our disposal gives us all the appearance of being more intelligent, we're making more and more silly mistakes, what I term constant partial stupidity.

I can read newspapers and websites from all over the world, at a time and a place of my choosing, and I can communicate with their authors, too. But I miss old-fashioned conversations and serendipitous encounters. Even when I do see people, the chances are that our chat will be fleeting or else the pudding of chilled berries will be interrupted by Apples and BlackBerries, at which point any interesting ideas will be frozen out.

But enough about me—over to you. Why, in an age of too much information and too little time, should you read this book?

Whether you want to find out about the benefits of boredom, mental privacy, the rise of the screenager, the sex life of ideas, or how digital objects and environments are changing our minds, you'll find thought-provoking discussion and practical suggestions about what's happening and what we can do about it. This is a book for anyone who's curious about rethinking their thinking or about unleashing the extraordinary creative potential of the human mind.

I am often described as a futurist, but my opinion about what might happen next is continually evolving. However, what I can say with some level of certainty is what is happening to me right now and where I am going next. So the book is about some of the emerging trends and "weak signals" I am watching and about what is going on inside my head.

⊓ The culture of rapid response plus ease of access to anything is encouraging mistakes. This is leading to a state of constant partial stupidity and multitasking mayhem. While multitasking means that we are getting better at thinking faster, the quality of that thinking is suffering. We can do more than one thing at once, but we can rarely do them well. Some studies suggest that multitasking increases stress-related hormones like adrenaline and cortisol and that this is prematurely ageing us through what is called biochemical friction. The backlash to multitasking will be a trend called single tasking, an idea borrowed from the Slow Food movement.

⊓ While screens offer us many opportunities, they can encourage thinking that is devoid of context, reflection, and an awareness of the big picture. Similarly, the trend for packaging information in byte-sized chunks means that we are sprinting toward the lowest common denominator. The countertrend to this (again an offshoot of Slow Food) will be "slow media"—long copy analysis and slow, paper-based communication.

⊓ We are living faster than we are thinking. We relish the speed of communication that is possible, but it is sometimes forcing us to respond without thinking things through properly. We need to slow some things down a little. A study at the Unconscious Lab at Radboud University in the Netherlands found that we make more effective decisions if we walk away from a problem and allow our brain to mull it over from a different perspective. We also need to step off the "speed is good" treadmill and deal with our fear that a slower pace will somehow have a negative impact on economic growth or progress.

⊓ While we benefit from the ubiquity of information and the possibilities of greater communication, constant digital disruptions and too much information are atomizing our attention and splintering our concentration. We are finding it difficult to remain focused and we are becoming addicted to the screen. We need to get away from the idea that all information is useful and adjust to a new reality where attention is power and it is trust in information that is critical.

◫ The constant flow of information on what other people are doing allows us to get a sense of their lives. Small bits of information, mundane and trivial though they may be on their own, eventually build into a kind of narrative. Scientists have called this phenomenon ambient intimacy, similar to how you can pick up another person's mood by being close to them and decoding the small signals they transmit. But constant connectivity means that we are replacing intimacy with familiarity, and this can also make our physical relationships with other people more ephemeral. Hence, we face a threat of widespread digital isolation. Expect to see a significant "return to the real," which will be linked to trends such as authenticity, localism, and craft.

◫ We have greater choice and more personalization, but concentrating on ourselves can reduce the opportunity for serendipitous encounters, with both people and information. We are shutting ourselves off from potentially valuable experiences and lessons. For example, work at the University of Chicago demonstrated that a more restricted range of sources are being cited in academic journals as sources move online, concentrating on fewer, more recent articles. Our thinking needs to be not merely deep but also wide, allowing for the cross-pollination of ideas and activities.

◫ The anonymity of the web is eroding empathy, encouraging antisocial behavior, and promoting virtual courage over real emotion. At the same time, oversharing information about our precise location or interests may let us know who else is in the vicinity, but it is also making us vulnerable to everyone from advertisers to burglars. Digital immortality also means that it is becoming increasingly difficult to forget previous actions or to get past our past.

◫ Online crowds are drowning out individual wisdom and experience. Online collectivism via social networks also means that we face pressure to be online and to conform to group norms. We are finding it much harder to escape from the presence of others and to be truly alone to concentrate on our thoughts and our ideas. What is known as attention restoration theory claims that just like people need to sleep, our brains need to take time out from the deluge of outside stimuli in order to relax and restore effective functioning. We rarely take pleasure in doing nothing and just enjoying our surroundings, yet doing so is one way of improving our mind and its capabilities.

PART ONE

HOW THE DIGITAL ERA IS CHANGING OUR MINDS

Chapter I

The Rise of the Screenager

"Computers are useless. They can only give you answers."
Pablo Picasso

Observe for a minute a teenager in their natural habitat, in front of a screen. Chances are that they're not speaking but furiously tapping a keyboard. They appear to be in a hurry and, one suspects, they aren't fully concentrating, waiting, as they undoubtedly are, for some new bit or byte of information to flash across the screen.

Today's teenagers are better described as "screenagers," a term popularized by Dan Bloom to describe the act of reading on screen. They're woken up by an alarm on a cellphone and they check the latest gossip on the same device, often before they get out of bed. They go to school or work in a vehicle that features screen-based information or an entertainment system, and they spend most of their day interacting with one kind of screen or another. In the evening they interact with their friends via screens and may finally sit down to relax with the internet. According to one 2009 study, an average of 2,272 text messages a month are currently sent or received via a US teen's phone screen; a 2010 report found that text messaging and social networking account for 64 percent of cellphone use among 16–24 year olds in the UK.

Don Tapscott, author of *Growing Up Digital*, claims that a student nowadays will have been exposed to 30,000 hours of digital information by the time they reach their 20s. Similarly, a Kaiser report that surveyed 8–18 year olds about their media consumption habits found that the total amount of leisure time American kids devote to media (most of it screen based and almost all of it digital) is "almost the equivalent of a full time job." Conversely, consumption of printed media, including books, is generally in decline.

Given the prevalence of screen culture among teens, what new kinds of attitudes and behaviors will we, as parents and employers, have to contend with?

10 ways screenagers are thinking differently

⌐ Screenagers prefer multitasking, parallel processing, and personalized experiences, read text in a nonlinear fashion, and prefer images over words.

⌐ Memory is something found on a hard drive. If they need information they Google it.

⌐ The ability to create, personalize, and distribute information easily is creating more of a focus on the self.

⌐ Screenagers frequently use digital devices to avoid confrontation and commitment.

⌐ Virtualization is removing the necessity for direct human contact and this is breeding a generation that prefers to deal with a machine than a human.

⌐ The reset generation thinks that if something goes wrong they can always press a button and start again.

⌐ The digital generation demands sensory-laden environments, instant response, and frequent praise and reward.

⌐ Screenagers live in the now and everything is just gr8, although they may be less literate and numerate than their forebears.

⌐ The screenage brain is hyper-alert to multiple streams of information, although attention and understanding can be shallow.

⌐ The screenage brain is agile but is often ignorant of wider context and culture.

They want it and they want it now

Screenagers have a desire for personalized experiences, and a preference for reading text in a nonlinear fashion and for images over words. They also want speed. They expect things to happen quickly and, as a result, have next to no patience. Digital content is usually available almost immediately and this mindset of instant digital gratification is translated to the nondigital world. Waiting 90 seconds for a hamburger is ridiculous to the average screenager. So too are queuing in a bank and physically interacting with someone you don't know.

A study by the Social Issues Research Centre (SIRC), a UK-based thinktank, says that Generation Y—usually defined as those born between 1980 and 1999—want it all and expect to get it. Along with instant gratification, the buzzwords of these digital natives are eclecticism and diversity. This is a sampling generation who expect to do whatever they want, whenever they want. This means buying single music tracks rather than whole albums or swapping jobs at the drop of a pay check. As one member of the SIRC study put it, "Nothing is out of our reach, we can get anything shipped from anywhere in the world in a couple of days."

People 35 and over use a cellphone to manage their day. For those under 35, and especially screenagers, a cellphone is a proximity device that allows them to reshape time and space. While mobile phones are replacing wallets, watches, and doorbells ("hello, I'm outside, can you let me in?") they also allow people not to commit, or to save commitment to the very last second. In the digital era it's impossible to be late because you simply reschedule. Commitments are similarly fluid because a better opportunity might always show up, hence everything's done at the last minute. Got somewhere to go? No need for a plan, or a map; just make it up on the run.

Personal communication is changing too. Want to dump your boyfriend? Just alter your profile status on Facebook from "in a relationship" to "single." After all, if you'd wanted to speak to him in person you'd have sent him a text ("eLoves me, eLoves me not"). If you need to communicate with someone it is usually unnecessary to see them physically. These are people (to paraphrase Nick Bilton, author of *I Live in the Future and Here's How It Works*) who do not see any distinction between real-life friendships that involve talking or looking someone in the eye and virtual ones, where communication is through email or text message. Facebook and Twitter as well as virtual communities such as Second Life also feed into the desire to be reassured that one is not alone, so screenagers use this kind of site to check on their own existence and coalesce around an ever-changing universe of friends and online culture.

This constant flow of information does indeed allow us to get a sense of other people's lives. Small bits of information, mundane and senseless though they may be on their own, eventually build into a kind of narrative. Scientists have called this phenomenon ambient awareness or ambient intimacy. It is similar to how you can pick up another person's mood by being close to them and decoding the small signals they transmit.

Life in a virtual world has benefits in the real world too. At Stanford University, the Virtual Human Interaction Laboratory (VHIL), run by Assistant Professor Jeremy Bailenson, researches how self-perception affects human behavior. Specifically, the lab studies how online activities influence real life. One of its findings is that there is a significant bleed between virtual experiences and real-life attitudes and behavior, in both directions. For example, if you become increasingly confident in a virtual world, this confidence spills over into the real world.

Nevertheless, we may also be developing a new generation who lack resilience and who believe that when things go wrong, all they have to do is press a button and things will quickly go back to the beginning for another attempt. If so, what will happen when something difficult shows up that they can't avoid, forward, or delete?

These might not sound like gargantuan changes, but what starts as a behavioral shift tends to flow into attitudinal change, which, in turn, becomes social change. Gen Y will be running the world in 10 or 20 years' time. They are the next wave of employees and if you are not working alongside them already, you will be in the not too distant future. You might also be wanting to know the best way to handle them as a teacher or parent.

Teachers may face a conflict of teaching and learning styles. Older teachers generally teach face to face and proceed in a logical or step-by-step basis. In contrast, younger students tend to jump around from one idea or thought to another and expect sensory-laden environments as a matter of course. They also want instant results and frequent rewards, whereas many teachers regard learning as slower and serious and consider that students should just keep quiet and listen. We are in for some rather stormy weather over the next few decades as the analogue minds of both teachers and parents clash with the attitudes and behaviors of digital minds. As David Levy, the author of *Scrolling Forward*, comments, there could be a "conflict between two different ways of working and two different understandings of how technology should be used to support that work."

In the world of work it's much the same story, although many of these issues have yet to surface in organizations because only recently have screenagers started full-time employment. Change will take longer to filter all the way through to the most senior levels in organizations. In the meantime, expect attracting and retaining talent to become much harder, especially if the economy is healthy. If those in the digital generation feel that they are not developing or progressing rapidly up the corporate ladder, they will simply leave. The idea of serving an apprenticeship within an organization is dead conceptually as well as economically.

CONNECTIVITY ADDICTION

We may be behaving like screenagers already, however. A study from the University of California (Irvine) claims that we last, on average, three minutes at work before something interrupts us. Another study from the UK Institute of Psychiatry claims that constant disruption has a greater effect on IQ than smoking marijuana. No wonder, then, that the all-time bestselling reprint from the *Harvard Business Review*, a management magazine, is an article about time management. But did anyone find the time to actually read it properly?

We have developed a culture of instant digital gratification in which there is always something to do—although, ironically, we never seem to be entirely satisfied with what we end up choosing. Think about the way people jump between songs on an iPod, barely able to listen to a single song, let alone a whole album. No wonder companies such as Motorola use phrases like "micro boredom" as an opportunity for product development.

Horrifyingly, a couple in South Korea recently allowed their small baby daughter to starve to death because they became obsessed with raising an "avatar child" in a virtual world called Prius Online. According to police reports, the pair, both unemployed, left their daughter home alone while they spent 12-hour sessions raising a virtual daughter called Anima from an internet café in a suburb of Seoul.

Internet addiction is not yet a globally recognized medical condition, but it is only a matter of time. Already 5–10 percent of internet users are "dependent," according to the Computer Addiction Center at Harvard's McLean Hospital. This is hardly surprising when you stop to consider what is going on. According to a University of California (San Diego) study, we consumed three times more information in

2008 than we did back in 1960. Furthermore, according to Clifford Nass, a professor of communications at Stanford University, there is a growing cohort of people for whom the merest hint of new information, or the faintest whiff that something new is going on somewhere else, is irresistible.

You can see the effect of connectivity cravings first hand when people rush to switch on their cellphones the second their plane lands, as though whatever information is held inside their phone is so important, or life threatening, that it can't wait for five or ten minutes until they are inside the airport terminal. I know. I do it myself.

The thought of leaving home without a cellphone is alarming to most people. So is turning one off at night (many people now don't) or on holiday. Indeed, dropping out of this hyper-connected world, even for a week, seems like an act of electronic eccentricity or digital defiance.

In one US study, only 3 out of 220 US students were able to turn their cellphones off for 72 hours. Another study, conducted by Professor Gayle Porter at Rutgers University, found that 50 percent of BlackBerry users would be "concerned" if they were parted from their digital device and 10 percent would be "devastated."

It's more or less the same story with email. Another piece of research, by Tripadvisor.com, found that 28 percent of respondents checked email at least daily when on a long weekend break and 39 percent said they did so at least once a day when on holiday for a week or more.

A study co-authored by Professor Nada Kakabadse at the University of Nottingham in the UK noted that the day might come when employees will sue employers who insist on 24/7 × 365 connection. Citing the example of the tobacco industry, the researchers noted how the law tends to evolve to "find harm." So if employers are creating a culture of constant connectedness and immediacy, responsibility for the ensuing societal costs may eventually shift from the individual to the organization. Broken marriage and feral kids? No problem, just sue your employer for the associated long-term expenses.

A banker acquaintance of mine once spent a day in a car park above a beach in Cornwall because it was the only spot in which he could make mobile contact with his office. His firm had a big deal on and his virtual presence was required. "Where would I have been without my BlackBerry?" he said to me later. My response was: "On holiday with your family taking a break from work and benefiting from the reflection that distance provides." He hasn't spoken to me since we had this conversation, although he does send me emails occasionally. I usually pretend that I'm on a beach and haven't received them.

It's happening everywhere. I have a middle-aged female friend (a journalist) who goes to bed with a small electronic device every night. Her husband is fed up and claims it's ruining their sex life. Her response is that she's in meetings all day and needs to take a laptop to bed to catch up with her emails. This is a bit extreme, but I know lots of other people who take their cellphones to bed. How long before they're snuggled up late at night "attending" meetings they missed earlier, having downloaded them onto their iPad or something similar? Talk about having more than two people in a marriage.

Our desire to be constantly connected clearly isn't limited to work. Twitter is a case in point. In theory, Twitter is a fun way to share information and keep in touch, but I'm starting to wonder whether it's possible to be *too* in touch. I have some friends who are "Twits" and if I wanted to I could find out what they're doing almost 24/7. One, at least, will be

"Eating marmite toast" at 7.08 pm and the other will be "In bed now" at 11.04 pm or "looking forward to the weekend" at 11.34 pm. Do I need to know this?

Why is all of this significant? In *A Mind of Its Own*, Cordelia Fine makes the point that the brain's default setting is to believe, largely because the brain is lazy and this is the easier, or more economical, position. However, when the brain is especially busy, it takes this to extremes and starts to believe things that it would ordinarily question or distrust. I'm sure you know where I'm going with this but in case you are especially busy—or on Twitter—let me spell it out.

Our decision-making abilities are at risk because we are too busy to consider alternatives properly or because our brains trip us up by fast-tracking new information. We become unable to exclude what is irrelevant and retain an objective view on our experience, and we start to suffer from what Fredric Jameson, a US cultural and political theorist, calls "culturally induced schizophrenia."

If we are very busy there is every chance that our brain will not listen to reason and we will end up supporting things that are dangerous or ideas that seek to do us, or others, harm. Fakery, insincerity, and big fat lies all prosper in a world that is too busy or distracted. Put bluntly, if we are all too busy and self-absorbed to notice or challenge things, then evil will win by default. Or, as Milan Kundera put it: "The struggle of man against power is the struggle of memory against forgetting."

Crikey. That sounds to me like quite a good reason to unsubscribe from a few email newsletters and turn the cellphone off once in a while—to become what Hal Crowther terms "blessedly disconnected." The future of the planet and life as we know it are clearly at stake.

Multitasking mayhem

Screenagers have a predilection for multitasking and parallel processing. An Ofcom report in 2010 found that while 16–24 year olds spend 6.5 hours a day on media and communication, 29 percent of that time they are multitasking, so squeezing in 9.5 hours' worth of activity. You can study, be on Facebook, watch television, and have a mobile phone clamped to your ear, but is anything of substance going into your brain? Academic and author Mark Bauerlein quotes an American student who says: "I can't concentrate on my homework without the TV on. The silence drives me crazy." Compare this with a study at UCLA, which found that multitasking had an adverse effect on learning. "Even if you learn while multitasking," says Russell Poldrack, co-author of the study, "that learning is less flexible and more specialized, so you cannot retrieve the information as easily... When distractions force you to pay less attention to what you are doing, you don't learn as well as if you had paid full attention."

Scientists using functional magnetic resonance imaging (fMRI) have discovered that the constant switching required to multitask effectively is damaging some of our higher-level brain functions, especially those related to memory and learning. We can just about cope with doing two things at once, but we often can't remember what we've done or how or why we did it. Some studies suggest that multitasking increases stress-related hormones like adrenaline and cortisol and this is prematurely ageing us through what's called biochemical friction.

Research by Eyal Ophir and others at Stanford University established that heavy multitasking students not only took longer to respond to tasks than light multitaskers, they made many more mistakes too. On the plus side, heavy multitaskers are among the first to glimpse new information, so they are good at spotting new ideas. The downside is an inability to focus on what they are doing, so they are somewhat flighty and flakey.

There is also what Edward Hallowell, author of *Crazy Busy*, refers to as "a constant low level of panic and guilt." These thoughts are echoed by a study from the University of California (Irvine), which found that people who are continually distracted by email suffer from higher levels of stress; and by Gary Small at the University of California (Los Angeles), who says that such stress can be linked to lower levels of short-term memory.

No wonder Bill Joy, co-founder of Sun Microsystems, likens the twenty-first-century teen to a village idiot. Speaking at the Aspen Institute's 2006 Festival, he said: "This all, for me, for high school students, sounds like a gigantic waste of time. If I was competing with the United States, I would love to have the students I'm competing with spending their time on this kind of crap."

Or, as authors William Strauss and Neil Howe have observed, "The twenty-first century teen, connected and multi-tasked, autonomous yet peer-mindful, marks no great leap forward in human intelligence or global thinking."

The screenage brain

Comparing the differences between generations is fraught with difficulty, but it is still one of the better ways to foresee the immediate future. Many compare analogue thinkers (Gen X and Boomers) to digital thinkers (Gen Y and Millennials), or digital immigrants to digital natives, terms coined by writer and games designer Marc Prensky. To quote Prensky: "Today's students have not just changed incrementally from those of the past... a really big discontinuity has taken place. One might even call it a 'singularity'—an event which changes things so fundamentally that there is absolutely no going back." This so-called singularity is the arrival and rapid dissemination of digital technology in the last decade of the twentieth century. Gen Y and younger have spent their whole lives surrounded by digitization and connectivity.

How will the digital generation think? What will it be like to grow up with smart, emotionally aware devices? Will the minds of future humans merge with machines to create some kind of posthuman hybrid? A new type of mind is emerging from an osmotic interaction with digital objects and environments. This digital mind is, in the words of Bauerlein, "mentally agile but culturally ignorant." It is "highly aware of itself and others in its immediate (digital) proximity but is impatient and surprisingly ignorant of the wider world."

Naomi Baron, a linguist at the American University in Washington DC, claims there is an "intellectual torpor" among today's students. Corner cutting (to save time) is prevalent and, more worryingly, students are thinking in incoherent snippets. This thought is echoed by Maryanne Wolf, a neuroscientist working at the Center for Reading and Language Research at Tufts University in the US. She claims that students skim, skitter, and are generally pointillistic in their approach to knowledge. When you can Google information, why do you need to form or remember knowledge?

For the last quarter of a millennium there has been a general consensus, certainly in western cultures, that words were important and that certain rules should be strictly observed. But spelling, syntax, and grammar no longer matter to screenagers. They're after speed and quantity of communication, hyper-alertness, search and seize.

The PC argument (that's Personal Computer) is that language has always been fluid and therefore we shouldn't worry too much about how people express themselves. After all, Chaucer couldn't spell in today's terms and it's really the thought that counts. It's all gr8.

And in the future, our interaction with machines will be predominantly oral and visual. We will ask machines questions and they will answer back. We will listen to literature and watch books. As science journalist James Gleick has observed: "We have learned a visual language made up of images and movements instead of words and symbols." But spelling, syntax, and grammar do still matter. So do physical books. They all convey ideas and if you restrict any of them you constrain thinking and discussion.

Worrying about the future of thinking is nothing new. In *The Shallows*, Nicholas Carr points out that in Plato's *Phaedrus*, Socrates laments the new reliance on the written word. Socrates believed that, while writing might give the superficial appearance of wisdom, this was at the cost of true insight. Equally, in the fifteenth century the humorist Hieronimo Squarciafico thought that Gutenberg and the relative ubiquity of printed books would render men lazy and "less studious."

Nevertheless, writing and printing did not reduce concentration, they increased it. Writing something down on paper (which in those days was expensive) or slowly reading a book (even more expensive) made you think hard and long. Reading was deep, deliberate, focused, and sustained as a result. Not any more. Technology such as cellphones, search engines, and email does the very opposite. It can create a cerebral whiteout. The writer Geordie Williamson referred to this as "another powerful accelerant in the outsourcing of the human mind." We may be reading more in terms of sheer volume, but most of this reading (and writing) is in short snippets. It is not the kind of reading or writing that I believe is important.

Compare physical books with their electronic equivalent. Digital books contribute to an accelerated pace where the acquisition of facts is almost devoid of broad understanding, narrative, or context. Physical books, in contrast, allow people to slow down and reflect. Physical books (and face-to-face conversation) shape and direct our thinking in ways that digital information does not.

Not only that, research by the University of Connecticut found that web users are consistently poor at judging whether web-based information is trustworthy or not. The study asked students to look at a spoof web page, http://zapatopi.net/treeoctopus, about a rare tree octopus. 90 percent of students thought the page was a trustworthy and reliable source, despite its mention of conservation organizations such as Greenpeas (saving the world from humans) and People for the Ethical Treatment of Pumpkins.

Another study, this time by web researcher Jacob Nielsen, tested 232 people reading material on screens using eye-tracking tools and found that only six participants read web pages linearly. Everyone else jumped around like caffeinated rabbits, briefly chewing on a bullet point and then careering off to snack on a patch of color or perhaps a typeface change. In a further project, Nielsen found that teens were faster than adults at reading online but their attention spans were much shorter, so anything difficult tended to be skipped.

The key point about digital or screen reading versus reading on paper is that books are part of a system of thinking. Books do not stand alone. They are contextual, both in relation to other books and to their historical setting. Digital books, and screen reading in general, are different because information becomes content, separated from any clarifying context. Any substantial move away from books would therefore represent a shift away from part of our cultural heritage and understanding.

Reading on a screen, especially one surrounded by hyperlinks, is fast and is suited to foraging for facts. In contrast, reading on paper is reflective and is best suited to trying to understand an overall argument or concept. Both forms of reading (both forms of technology) ought to be able to live alongside each other.

We need to get the balance right and restrain the use of certain devices in certain environments. Pencils and books are still important and employers and educators need to think much more clearly before they plug in and log on. In some instances we need to decelerate learning. After all, we've had slow food and slow cities, so why not slow learning and slow thinking?

Are IQ tests making kids stupid?

The number of adults with a university-level education in the UK has increased markedly over the past 40 years. According to OECD research, 29 percent of people aged 25–34 have what the OECD calls a Type A tertiary education (essentially from universities, which have expanded to allow greater access), while for those aged 55–64 this figure is just 16 percent. In short, more people are now going to university—well done dons. But while a numerical expansion of academia means more thinking, this doesn't necessarily translate into better-quality thinking. Moreover, for every study or statistic "proving" that we are getting smarter in a general or overall sense, there seems to be another "proving" the very opposite, especially when one starts to look at very specific areas or traits.

For example, a report from the US National Endowment for the Arts has linked flat or declining national reading scores with a reduction in the number of teens reading books. In 1984 around 30 percent of 17 year olds claimed to read almost every day for pleasure. By 2004 this figure had fallen to around 20 percent and 19 percent also said they never read, up from 9 percent in 1984. In 2006, the Conference Board in Canada found that almost 90 percent of employers thought that "reading comprehension" was "very important." It seems there is a gap between the behavior of students and the needs of employers.

And here's an interesting conundrum. According to James Flynn, Professor Emeritus at the University of Otago in New Zealand, IQ scores rose steadily throughout the twentieth century. Indeed, there has been a fairly consistent three-point increase every decade since testing was first introduced. Furthermore, this smartening up is accelerating. The annual increase between 1947 and 1972 was 0.31 points but during the 1990s the annual increase was 0.36 points. In between watching *Woody Woodpecker* (1947) and *Harry Potter and the Philosopher's*

Stone (2001) kids got smarter and are now far smarter than their grandparents, who were brought up reading difficult books and writing out long essays by hand.

Yet how, in an age of failing schools, illiteracy, reality television, *Big Brother*, Paris Hilton, and Guitar Hero can we possibly be getting any smarter? There are lots of theories, such as better diet, more leisure time, smaller families, and even video games. But perhaps a better explanation lies in how we view intelligence, and especially how we measure it via the IQ tests themselves.

An IQ test measures general intelligence and various commentators (most notably Stephen Jay Gould and Howard Gardner) have thrown large spanners into how such a test works. IQ tests are a reasonable measure of logical problem-solving ability or abstract reasoning. People are asked to find similarities between different objects, arrange items into logical sequences, or mentally shift geometric shapes. These are tests that involve classification and correlation, and it appears that these abilities have been rising consistently for a very long time. But this has nothing whatsoever to do with video games or anything connected to the internet. This is hardly surprising. In 1900, most people would have been employed in either agriculture or factories and there was limited demand for abstract or conceptual thinkers. Outside of universities, "intelligence" would have been seen as something relating to practical (physical) problem solving. Hence intelligence—as measured by an IQ test—would have been low. But over the last 100 years or so, the requirement for abstract or conceptual thinking has grown substantially. As a result, we have got much better at precisely the kind of nonverbal intelligence that IQ tests measure best.

This can help to explain why IQ scores are increasing while at the same time millions of workers in the UK and the US are functionally illiterate and innumerate. It also sheds light on why screenagers get good IQ test scores while at the same time the number one gripe from employers is a lack of basic reading, writing, and arithmetic skills.

If you split out test results for literacy and numeracy, you will find that students today are no smarter—and in some cases they are considerably more stupid—than students 50 or 60 years ago. According to one Australian study, a teenage student in 2003 was about a quarter of a grade level below a comparable student in 1964. Or, as Nicholas Carr points out, tests that measure "memorization, vocabulary, general knowledge, and even basic arithmetic have shown little or no improvement."

In other words, cognitive skills are rising dramatically and this is connected to improved use of logic and hypothesis. Put simply, we are living in a time where moral and ethical questions are more important and where both work and leisure are more intellectually than physically challenging.

This is something the writer Steven Johnson picked up on in his provocative book *Everything Bad Is Good for You*. He has argued that popular culture (everything from episodes of *Law & Order* and *CSI* on television to video games such as Tetris, Myst, and Grand Theft Auto) is more intellectually demanding and individuals are being forced to work a bit harder mentally to participate. Our external environment is changing and this is having a positive effect on screenagers' abstract reasoning abilities. Our minds, after all, are wired to adapt.

We do need minds that can react instantly or monitor fast-flowing information. We do need minds that can screen. In this sense, screenagers have an advantage. But we also, crucially, need curious, playful, imaginative, deep minds that can dream up big ideas and express them crisply and succinctly to other people in a compelling manner. And this, dear reader, takes us on to the next chapter—straight back to school and to the generation coming next.

Chapter 2

Pre-Teens: An Apple for Every Teacher

"The mind is not a vessel to be filled but a fire to be kindled."
Plutarch, 46–120AD

It is the current generation of under-10s (Millennials, Gen A, Gen Z, iGen, or iKidz) even more than screenagers who will really push the cultural and neurological boundaries. For instance, with younger children it's sometimes not clear whether they're in the real world or a virtual world and whether they perceive any difference between the two. This is the generation for whom there will be much less distinction between biological objects and machines, and for whom genetics, robotics, nanotechnology, and the internet will evolve beyond anything any adult can currently imagine in their wildest dreams.

Technology is deeply embedded in their lives. Most kids aged around 5 know what Google is (it's a verb) and they know how to use it. A report in the *Korean Herald* claims that 52 percent of Korean infants aged 3–5 regularly use the internet, spending an average of four hours per week online.

In the UK, according to an Ofcom report, 80 percent of children have a cellphone by their 11th birthday and by the age of 12 they're spending 8 hours a week online. Between the ages of 10 and 11, the average British child spends 900 hours at school, 1,300 hours with their family, and 2,000 hours in front of a screen. Similarly, during a typical week, an American kid will spend 10 hours on a computer, 6 hours on the internet, and 13 hours watching television.

Never before have kids around the world been so connected to each other and to a much wider community at a much younger age (primarily through the internet and cellphones). What children get taught at school, and the values they are given by their parents outside school, is diluted by the information and attitudes disseminated by technology, especially by mobile devices over which parents and teachers can exercise very little control.

Common sense would suggest that being glued to a screen all the time isn't a good idea. But can digital stimuli change the architecture of the brain? A growing number of people think they can. For example, Nicholas Carr has commented, "With the exception of alphabets and number systems, the Net may well be the single most powerful mind-altering technology that has ever come into general use... the possibility of intellectual decay is inherent in the malleability of our brains."

According to Dimitri Christakis, a pediatrics researcher at the University of Washington, every hour of television a child watches before the age of 4 results in a 9 percent rise in the risk of attention deficit problems by the age of 7. To put some context behind that statement, 79 percent of children in Britain have a television in their bedroom and prescriptions for Ritalin to treat hyperactivity have grown by 300 percent over the last decade. In the US, 33 percent of children live in a home in which the television is on "always" or "most of the time," and there is even research claiming that 54 percent of children aged 4–6 would prefer to watch television than spend time with their fathers. Add in the internet and electronic gizmos and children are continually engaging with toys and other experiences that offer an instant reward. They push a button and something happens. As a consequence, they crave constant change and are often overstimulated. They expect to be entertained and visually stimulated (both at home and at school) and they are losing the age-old ability to entertain themselves quietly or just think.

It is worth dwelling on the young brain. Back in 1981, David Hubel and Torsten Wiesel won a Nobel Prize for the discovery that there are windows of time in the young brain when major circuitry deep inside our heads is laid down. These windows, where visually based information is fed through to the brain, can be open for just a few weeks. How do we know this is true? Because there are children whose eyesight has been seriously damaged during these critical periods. Research has established that when certain parts of our wiring system are not put in, certain areas of our brains go unclaimed as a result. Usually this wouldn't be a problem because the brain rewires itself throughout its life, but it now appears that, in a few unfortunate cases, an electrician called on the brain as arranged but there was nobody at home. And you can't get a second appointment. That's why we need to give serious consideration to the effect of current developments on young brains.

Not enough thinking inside the sand box

Free play—the term used to describe play that is unstructured, imaginative, and initiated by the child itself—is in decline, according to a study in the *Archives of Pediatrics and Adolescent Medicine*. Part of the explanation is the rise in interactive toys and games that follow a priori rules (electronic toys and some video games have formal rules established in advance by their makers). In free play there are no rules and kids are forced to bring to bear their imagination and curiosity.

For example, a generation ago kids built model aircraft from kits and made things out of Lego. These toys actively engaged their minds and developed their imaginations and open-ended problem-solving abilities, especially when you could subvert the instructions or hack the parts to build something else. These days Lego bricks come in preformulated kits (buying only the bricks is quite difficult) and once a kit is assembled there's not much else you can build, because there are too many custom parts that have only one use.

I was putting one of my sons to bed recently and I showed him the *Eagle Book of Hobbies* from 1958 (a bit like a better version of *The Dangerous Book for Boys*). He was blown away by it. Inside were instructions about how to build model aircraft from scratch, as well as chapters on camping, metalwork, small bore rifle shooting, and chemistry (make your own chlorine gas!). We are now so paranoid and risk averse that most of the fun bits have been removed from children's lives.

Fantasy games that are wholly invented by the child are far from pointless either. For example, a HighScope Educational Research Foundation study found that children from poor backgrounds who went to play-oriented pre-schools were more socially adept in later life than similar children who went to play-free schools where strict academic instruction was the norm. By age 23, around one third of the individuals missing out on play had been arrested compared to less than 10 percent of the other group. A study by Pellegrini argues that play fighting among boys improves problem solving; a study by Panksepp and others found that nonstructured play encourages the development of higher brain functionality (especially areas associated with emotions and socialization); and a study in the *Journal of Child Psychology and Psychiatry* found positive correlations between free play and emotional health. If you want evidence to support the argument that low-tech and no-tech play is a good thing, it's around in spades.

Free play among children is not dissimilar to the kind of play seen in young animals, which suggests that it could have deep evolutionary origins. Interestingly, while animal play is also voluntary, it appears that animals will not play if they become stressed or hungry, so perhaps there is a parallel here with children from poor or abusive backgrounds. The 1990s example of thousands of young children locked away in Romania's 600 state institutions is a salutary reminder of the effects of such deprivation.

Outside such extreme examples, not engaging in any free play whatsoever is, of course, practically impossible. But what happens if children simply play less outdoors? According to the *Daily Mail*, 70 percent of British parents no longer allow their children to play outside unsupervised, and a survey by ICM for the BBC found that 49 percent of British children have now been banned by their parents from climbing trees. Within the space of a generation we have moved from saying "Go outside and play and don't come back until it's dark" to "If you don't behave you can't stay inside and use the computer." Well-meaning parents are worried about stranger danger and broken bones and a host of other childhood risks. But what are the effects of all this cosseting on young minds?

Another important factor is that free play should sometimes involve other children. Peer interaction is vital if children are to develop language abilities. Fantasy games involving other children are especially good because they force kids to react to things that only exist in other children's imaginations. Language and communication are extended via group play.

Whether any of this applies to fantasy games played in virtual worlds is an intriguing question. I suspect there are advantages to playing games like The Sims and Spore (it's been found that surgeons who play video games for a few hours every week make fewer mistakes in the operating theater than those that don't), but even these games are formulaic. They possibly quicken spatial intelligence and facilitate collaborative problem-solving skills, but they appear to do next to nothing for high-order thinking skills. Furthermore, even the very best games are still largely pre-determined compared to games that are purely played inside one's head.

The great irony here, of course, is that parents shy away from free play in favor of organized school or career-oriented activities because they believe these will make their kids smarter and more successful. Play is play and work is work, they assume. But a seminal study in *Developmental Psychology* found that kids who were allowed to play randomly with a range of common objects were much better at coming up with nonstandard uses for another object than groups of kids who were asked to do a range of other activities. In other words, free play fosters curiosity, imagination, and original thinking, which are precisely the traits that innovative companies say they are after these days.

The modern era is hugely complex and highly unpredictable. Change is endemic and we are often faced with things that need to be radically rethought. Moreover, individuals and organizations need to cope with unexpected situations and unfamiliar environments. If, as children, we engage in random, unstructured activities, we will develop better coping mechanisms. We will be more prepared to look at situations in fresh ways and to come up with novel solutions to open-ended problems. To paraphrase Shona Brown and Kathleen Eisenhardt, authors of *Competing on the Edge*, we must thus create a balance between structure and chaos, experimentation and execution, strategy and improvisation, if we are to reach our full potential.

Another worrying trend relating to young children is homework. In senior school the case for homework is watertight, but at junior school there seems to be little evidence that homework contributes in any meaningful way to development. So why do we endorse it? Parents may mistakenly believe that homework will give their children an advantage that ultimately translates into a long and happy life.

However, when very young children have excessive homework, parents have less time to spend with their kids—the average amount of time a father spends alone, one to one with his child, is 10 minutes per day—and, most critically, kids have less time for unstructured free play. As Albert Einstein said: "The intuitive mind is a sacred gift and the rational mind is a faithful servant. We have created a society that honors the servant and has forgotten the gift."

We have also created a society in which schools teach children how to pass exams but they don't generally teach children how to think. Happily, this isn't true of all schools. The primary school ranked #1 in Britain in 2009 (Combe Church of England Primary School in Oxfordshire) does both. It has classes in philosophy where under 11 year olds discuss whether big things cause happiness and school assemblies where children ponder whether rich is heavier than poor. But schools like this are the exception.

It is also well known that the more a parent talks to their young child, the faster the child's own language develops. The amount of attention parents give their small children is critical. Before the 1980s most prams faced toward the parent; nowadays prams and strollers tend to face away. Moreover, if parents are constantly connected to iPods and BlackBerries rather than their children, what does that do to the quality of their communication? It's not just babies who can be affected either. Older children consistently claim that they feel hurt when their parents' gaze turns toward electronic devices, especially during mealtimes, school pickups, sports events, or other extracurricular activities.

A cut-and-paste education

You'd think that schools would be one of the few places where digital technology would be managed or resisted, especially with younger students. The view might be that the physical relationship between a teacher and his or her (usually her in primary education) students is paramount and that technology should be introduced slowly and only when there is a proven benefit. You'd think that most parents would feel this way too. But the exact opposite seems to be the case. This is despite the fact that there is plenty of evidence to suggest that computers should be used with caution in certain educational contexts.

You can buy a lot of books for the cost of a single interactive whiteboard, but this doesn't seem to be stopping junior school classrooms turning themselves into the educational equivalent of NASA mission control. (An Apple laptop for every teacher and every child.) For example, the Australian government recently promised to spend AUD$2 billion to give a computer to every Australian student in years 9–12, while in the US the State of California is replacing all school textbooks with e-books. Nevertheless, I am not aware of any research that has found a positive correlation between increased expenditure on computers and improved student learning or exam results.

Quite the contrary, in fact. The Scottish Inspectorate of Schools writes: "Inspectors found no evidence of increased attainment in formal qualification or against nationally defined levels that could be directly attributed to the use of ICT in learning and teaching." Another report from Munich University looked at student assessment tests for 15 year olds across 31 countries and found: "Once other features of student, family and school background are held constant, computer availability at home shows a strong statistically negative relationship to math and reading performance and computer availability at school is unrelated to performance." One 2010 report co-authored by Professor Jacob Vigdor at Duke University in the US is even more damning. This study looked at the reading and mathematics scores of 15,000 children between 2000 and 2005, splitting out the scores before and after they had access to computers. The conclusion? Children with computers at home perform worse in exams than children without. And this study was conducted *before* Facebook and Twitter showed up on the screen.

ICT is not a silver bullet and on its own cannot magically turn a bad teacher (or school) into a good one. It's probably not a vote winner, but governments would probably get a much better result by hiring more teachers or paying teachers more money. In Australia, AUD\$2 billion roughly equates to an extra AUD\$8,000 per teacher per year, or a flood of new teachers or classroom assistants for 12 months.

It's how you use this technology (and when you don't) that counts. Using technology to retain the shortening attention spans of young children seems rather illogical to me. There is undoubtedly pressure to keep kids continually entertained, but if they're unsettled and unfocused, you need to slow things down, not speed them up. We're also becoming too focused on virtual interaction. If this technology is so great why, in the UK, are cognitive tests for 11 year olds showing scores that are, on average, two or three years lower than children of the same age 15 years ago?

Why books still matter

How young people are absorbing information is changing rapidly and radically. Passive media such as newspapers and books are now generally rejected in favor of interactive media and, ideally, media that can be co-created and controlled.

However, if we are personalizing everything, eventually there will be no shared experiences left. Young kids live very much in the here and now; both the past and the future are too far away. One potential consequence of this is that there will be a declining appreciation of physical books as historical artifacts—or of the value of older people because they represent a historical record of societal thinking, most of which is not available on Google.

Reading, especially reading fiction for pleasure, has been in decline for decades in most countries. This is a serious issue because reading has a direct impact on intelligence and attainment. For instance, a Kaiser report in the US called "Generation M" found that "leisure reading of any kind correlates more closely with a student's grades than any other media."

Books will survive in one form or another, but what are the implications for society when silent text is constantly under threat from restless hypertext? What happens to our culture when audio and video are deemed more useful than the written word? The answer may well be a postliterate society where most people can read but most choose not to. Words become art supplies. Visual thinking is the hallmark and feeds a mindless digital hedonism in which novelty is more important than quality.

Reading online and reading from a physical book are entirely different experiences. People are generally in a rush online and want to extract information or "value" as quickly as possible. Offline, in contrast, we tend to have a calmer mindset and, as a result, our imagination is more engaged. Empathy and clear communication also suffer when time is of the essence. The critical point here is distraction. Online we are continually tempted by hyperlinks to other information; on paper we are not. A halfway house is ebooks that are not permanently connected to the internet, although we need more research to establish whether the screen itself is having some impact.

A survey by the Association of Teachers and Lecturers in the UK claims that 25 percent of work handed in by pupils contains material copied directly from the internet. Sometimes this is done carefully; often it is not. One teacher responding to the survey mentioned a case where material had been copied so quickly that the student had not even thought to remove the web pages' ads. It is becoming so easy to copy things that some people are no longer thinking deeply about what they are doing. This isn't necessarily a problem. One teacher in the study remarked that most students who cut and paste material do so in ignorance. Once it is explained to them what cheating is and why it's a bad idea, most stop. However, the temptation will always be there.

As psychologist Robert Ornstein says: "Literacy is one of many means to knowledge." Life is getting faster and there's more information to absorb, so we must learn to find whatever it is we need quickly. A key skill is therefore knowing where to look for things and then rapidly analyzing what we find to extract what we want from the exaflood of information that's instantly available.

Skilled search is undoubtedly valuable in a world that is becoming choked with information. But spot knowledge (just-in-time learning) makes our understanding superficial and our obsession with speed will make mistakes more common in the future. Moreover, trying to absorb multiple sources of information at the same time forces our thinking to become shallower. There is little or no deep thinking and next to no reflection going on. There just isn't the time. According to a 2009 study by Eyal Ophir and Anthony Wagner in the US, multitasking is becoming an increasingly normal state. Younger people especially believe they are good at it, although the evidence suggests the opposite—they are among the worst. Furthermore, the study indicates that multitasking may harm thinking ability and we may be developing a mindset that is incapable of ignoring irrelevant information.

Does our lack of concentration (or our perceived lack of time) mean we have to create shortened versions of absolutely everything or put open-source textbooks online, adding some interactive, eye-candy element to keep the class awake? If kids find it difficult to concentrate or to "get" someone like Shakespeare, then making learning more interactive (more like Disneyland) is not always the right answer.

First of all, what's the problem with confronting children with a few things that are hard? Difficulty breeds resilience and this is a trait that is seriously lacking nowadays. In a nutshell, we have become too soft and our brains and bodies are accepting mush. We are shielded from difficulty and disharmony by concepts such as fairness and access. For example, hundreds of schools in Britain have now banned the use of red ink by teachers because the color is thought to be too "confrontational" or "threatening."

If we make education—and life—too easy we are in danger of breeding a generation that lacks critical life skills. Kids look sophisticated and worldly wise on the surface, but their confidence is becoming extremely brittle. Hard books should therefore be compulsory and, in the spirit of lifelong learning, there should be campaigns aimed at adults encouraging them to read more and to read better books.

Indeed, why not tax products and experiences that are known to provide no mental nourishment and use the revenue generated to subsidise other, more useful, mental experiences? This happens already with the arts (and the BBC historically) but I see no reason why we shouldn't extend the principle and think carefully about the implications of cerebral snacking.

In the future machines will be better than people at storing and applying information. Hence, educating individuals in just-in-time knowledge retrieval is a recipe for eventual oblivion or at least irrelevance. Once students have mastered the basics, we should be training them to think deeply and differently. But how can we achieve this if certain types of book migrate to digital formats? How do you instill rigor in a young mind that is being encouraged to jump from one picture to another on a small screen?

We need more childlike thinking

Children appear to be more "open" to new ideas than adults, but whether this is related to less experience or developmental differences between the two hemispheres of the brain is unclear. As children get older, their natural originality and sense of wonder are tempered, and eventually smothered, by socialization. Is this process of socialization being accelerated by digital technology? If it is, we could be in trouble, because youthful minds are incredibly valuable.

In primary and secondary education, we are generally taught that there are right and wrong answers to everything. With very few exceptions, we are taught that precise, logical thinking is the way to pass exams and gain university admission. Education focuses on sequential, reductive, and logical thinking. The system is single minded when it should perhaps be more broad minded and more systems minded.

As the former educationalist Ken Robinson has pointed out, the education system is created to meet the demands of a certain industry: further education. Hence certain subjects are deemed more practical or useful than others. But, as Robinson says, why shouldn't creativity be given the same status as chemistry? Moreover, we educate kids from the feet up. We start with the feet (walking), move on to the hands (writing), and then swiftly move up to the head (memory and, hopefully, understanding). However, this journey is biased toward, and rewards, only one half of the brain.

This may not be an issue if your aim is to grow logical, linear thinkers. But if as a society your aim is to foster individuals who can understand complex systems and creatively rearrange individual components, so they can ignite an innovation economy, then you need broad context as well as text. And in this context, inexperience and open minds can be quite valuable.

Knowingly or unknowingly, children and adults alike use mental filters to judge whether things are useful to us or not. But these filters can screen out new ideas too early. We become too concerned with the negative aspects of ideas or worry about what other people will think of them or us. This is problematic because the generation and propagation of ideas are inherently social.

Historically, young children didn't usually have an awareness of what other people thought of them. They were open-minded and were not hidebound by the potential consequences of their actions. They were naïve thinkers who would guess and experiment, unencumbered by the potential consequences. They were not frightened of being wrong either, often because they could make mistakes away from the penetrating gaze of adults and other children alike. They would also take their time to get things right.

These have always been traits that, unfortunately, are educated out of us as we grow older. As soon as we reach early adolescence we become focused on speed and what our peers might think of us. Hence, in school and at work, making a mistake is often the worst thing that you can do. This is only made more evident by constant connectivity. On sites such as Facebook people are continually checking up on each other. Even if you do choose to drop out of social networks, this doesn't mean that you disappear. People will still post images and comments about you, which is why most people choose to stay connected and manage what's going on. As someone commented to me: "If you don't dive in, other people will define you." If you make a mistake it's now very difficult to erase.

Nevertheless, making mistakes is absolutely essential for discovery, insight, and invention. Indeed, you usually learn more by getting things wrong than by getting things right. But here's the real problem. If the trend is toward kids growing up more quickly and the digital era is eroding privacy, what will happen to such childlike experimentation and innocent thinking?

A lack of experience was something that Seymour Cray (an early designer of high-speed computers) seized on. He had a policy of hiring young, fresh-faced engineers because they didn't yet know what couldn't be done. In a similar vein, Toyota once put together a "board" of children to advise on product development, while Hasbro has done much the same with children and toys. Even Xerox's famous Palo Alto Research Center (PARC) once asked school kids to attend a series of meetings on the future of technology.

Young people tend to have high levels of energy and confidence. They are also outsiders and have less respect for tradition or orthodoxy. This lack of experience (a lack of information cemented in their brains, if you like) can be a major negative, but it can be very positive too. They are virgin thinkers. This thought is similar to Einstein's point: "You can never solve a problem on the level on which it was created." Get too specialized, or too close, and you won't see anything. But step back and you sometimes see a new or larger picture.

Dare to dream

Language is largely developed during early childhood and is usually thought of as a product of the conscious mind. We eventually teach children about words and grammar, but in the early years much of what children learn is not formally taught. This links to new theories, such as that put forward by Michael Shadlen, a neurobiologist at the University of Washington, who has suggested that subconscious thinking (or what some scientists prefer to call unconscious, pre-conscious, or nonconscious) is not separate from conscious thought and is not unthinking or undirected either.

His theory is that our subconscious mind is always working and is continually scanning the external environment for tiny pieces of information (often things seen, heard, touched, or felt for less than a second) and then deciding whether or not to connect this information to the conscious mind. It is, in effect, a kind of gatekeeper and could play a vital role in memory formation, language development, and idea creation. Indeed, you could make the case that, at a conscious level, we are little more than machines unknowingly guided by something we do not even perceive is there.

If machines are becoming better at logical thinking, then surely schools should be creating the environment and experiences where more conceptual and emotionally based thinking will surface. Schools should be equipping children for a world where knowledge is no longer fixed and where value is primarily derived from the ability to connect emotionally with other human beings and coming up with practical (and sometimes thoroughly impractical) new ideas. We should therefore stop filling up kids' heads with quick facts divorced from a broad context and place kindling and light fires under their imaginations instead.

The purpose of education (especially in the early years) should be to instill and nurture a sense of wonder and natural curiosity. It is also about developing a humane mind. Similarly, the point of academics is not what *they* think but how they change the way that *you* think. We should therefore slow education down a little and become less obsessed with visceral effects. We need fewer facts and more context. We should become less obsessed with short-term measurement and less concerned with goals and "outcomes" too.

We also have to allow children to be childlike. They need to dream, unfettered by time, practical considerations, or dissent. If we don't do this, then we will create a generation with the means to find quick answers but no idea how to generate and explore deep and meaningful questions. Constant curiosity, together with dogged determination and time, seem to me to be at the heart of imagination and originality, but we do not currently place nearly enough value on these traits.

The bottom line here is that much of early education in particular—and life in general—has become risk averse and timetabled to produce an end product that is acceptable to society and malleable for business. If you stray from the track, it's increasingly difficult to get back on board and the range of skills that society considers acceptable is shrinking fast. Deep thinking is increasingly seen as elitist and we are witnessing dumbing down on a massive scale. Discovery, insight, and invention all require experimentation, which is often messy and unpredictable and requires a challenge to the status quo (authority), which may not be something the average teacher is happy to facilitate in the classroom.

Since the world is going to become faster, more technological, more virtual, and more prescribed in the future, then what kids—and the rest of us—need is a counterbalance. We need more low-tech freedom and experimentation. We need to tinker more. We need more blank sheets of paper and more pencils. We need to focus on the things that make us uniquely human.

10 ways our education system could stimulate young minds

- Focus on free play rather than organized and structured play to avoid hindering imagination.
- Don't place too much emphasis on narrowly measured outcomes.
- Avoid becoming obsessed with ICT at the expense of teachers and books.
- Emphasize creative as well as rational subjects to foster the original and imaginative thinking that companies are looking for.
- Reduce homework to increase family time and encourage imaginative free play.
- Find ways to help busy parents spend more time with their children.
- Encourage parents to set more boundaries and allow less virtual interaction.
- End the health & safety hysteria that keeps children indoors and limits physical exploration.
- Remember that all reading is not necessarily good and pay attention to the medium.
- Avoid letting information and attitudes from electronic devices, especially cellphones, dilute lessons and values from parents and teachers.

PART TWO

WHY THIS MATTERS

Chapter 3
Thinking About Thinking

*"In your heads must still be the remnant of a brain. In your
hearts must still be the desire to be a human being again."*
Patrick McGoohan, *The Prisoner*

It has become fashionable to argue that the box of bits known as
the human brain is like a complex machine. Specifically, the
brain is like a computer and the mind is its software or operating
system. Since the internet (supposedly a kind of brain that links
people and ideas) is every year doubling in size, it will soon sur-
pass human intelligence and eventually become self-aware. This
will inevitably consign human beings to the wastebasket of history. If this happens it won't happen overnight, but some people
are already placing bets as to when it will. The futurist and
inventor Ray Kurzweil believes that by 2019, we will be able to
buy a box with the processing power equivalent to a human
brain for around US$1,000, while Mitchell Kapor (the founder
of Lotus software) has predicted that a machine will pass the
Turing Test for artificial intelligence (AI) by the year 2029.

These ideas are not new. In 1938 HG Wells wrote a book of
essays called *World Brain* about a global encyclopedia that would
help every person on the planet become better informed. More
recently we've had Pierre Teilhard de Chardin's *Noosphere* and
ideas such as the hive mind and the semantic web. As for robotic
thinkers being a serious threat to humankind, this idea isn't new
either. But while all of these ideas make great newspaper head-
lines, they fundamentally misrepresent how the human mind
works and what makes it so special.

The human brain is extremely complex and possesses prop-
erties that most computer scientists and software engineers go to

great lengths to avoid. In other words, the human mind is far more complex than a simple series of zeros and ones. Currently a 5-year-old child has more common sense than a computer. Indeed, if Groucho Marx were alive today and researching AI, he would no doubt be repeating his famous line: "A child of 5 would understand this. Send someone to fetch a child of 5."

It's the same story with robots. My younger son was studying robots at school a while ago and I had a chance to look at some of the work in the classroom. On a piece of paper outlining the differences between humans and robots, one child had written, "Robots don't work." He was referring to general uselessness rather than a lack of suitable employment opportunities, and he wasn't totally wrong.

The reason that a child has more general intelligence than a computer (or robot) is that he or she has more experience of the outside world. As Neil Gershenfield points out in his book *When Things Start to Think*, the mistake of many AI researchers has been to spend too much time and money on computer brains and not nearly enough on computer bodies (i.e., senses). Cognition is linked to perception and the brain is a two-way street in terms of information coming in and going out. Likewise, wisdom (as opposed to intelligence) is connected to experience. If you want to make an intelligent machine, the first thing you need to do is design one with experience as well as reasoning.

And therein lies the uniqueness and value of the human mind, which means that we are in all probability years away from even coming close to true machine intelligence. On one of my favorite websites—Longbets.org, a site dedicated to improving long-term thinking—Nova Spivack, the co-founder of EarthWeb, is betting that by 2050, no synthetic computer or machine intelligence will have become truly self-aware (conscious). Moreover Moore's Law, the idea that computer processing power doubles every couple of years, has one fatal flaw, which is that energy consumption also increases. So while it may be possible to design supersmart machines, we may find out that we cannot afford to run them. Critics, of course, will point to events such as the IBM computer called Big Blue beating Gary Kasparov at chess in 1997. But while the computer was undoubtedly very good at chess, it was fairly useless at everything else. After all, if Big Blue had been truly intelligent, surely it would have been hanging around the bar afterwards with some other smart machines celebrating its win. Computers are also easily broken, at which point they don't work. While the human brain, in contrast, can be seriously damaged it generally still works, albeit in a different way. What is fantastic about the human mind is that while we have great difficulty remembering a sequence of more than seven or eight digits (in computer terms all this requires is 40 bits of memory, roughly what you'd find inside a $5 calculator), we can paint pictures or compose music that can make people cry or write books that can make people laugh.

10 ways our minds are different to our machines

- The basis of human intelligence is experience and is based on sensory awareness of information coming in as well as our response to it.
- Machines cannot think about their own thinking—they are not self-aware.
- Human beings possess generalized intelligence; machines are programmed for specific tasks.
- A machine lacks true senses—it can "know" it is cold, but it cannot "feel" cold.
- Machines do not have empathy or morality and they cannot feel love, joy, hate, or any other emotion.
- Electronic devices are not capable of creativity, intuition, or imagination.

⊐ People currently have mental privacy, but the workings of machines are transparent.

⊐ We can download information into a machine, but not yet into the human brain.

⊐ Machines do not possess a subconscious mind, yet this, more than the conscious mind, is the basis of most human thought and behavior.

⊐ The human brain has evolved over thousands of years so is highly resilient and adaptive to changing circumstances.

Computers are stupid

There are indeed parallels between human and machine intelligence in terms of information processing, but there the similarity ends. In short, computers cannot think about their own thinking. They are cold, calculating, and stupid. What computers *are* good at is logical analysis, but they are generally unable to criticize their own thinking, or create their own problems. Without consciousness a machine cannot be considered intelligent.

Even putting this slippery issue to the side, scientists can't even agree on what an artificial brain would represent. A human brain is not only information, or at least it's not information as we know it. Our brains are complex cauldrons of connections, chemicals, and electricity that, somewhat ironically, our conscious brains cannot understand. Get your head around that.

In a sense, the question of whether machines will ever be more intelligent than humans is misleading. We are likely to build specialized machines that will be much better than humans at certain routine, repetitive, or logical tasks, but it's unlikely that we will build general intelligence—a machine that can outperform humans across the board—any time soon. A big question is therefore not whether machines will get smarter and more human, but whether humans will get more ignorant and machine-like.

Neuroscience is still in its infancy. While there are massive insights and discoveries yet to be made that could turn current thinking upside down, it appears that networks of neurons are very different from networks of transistors. For example, the human brain and mind are interdependent, so the analogy with computer hardware and software is incorrect. Moreover, while ever more intelligent computers will be built, how will they ever be capable of doing many of the simple things that we humans take for granted?

Human thought is predicated on experience and a sense of self. While this is "felt" directly (via our senses), it is also taught to us by other human beings. In other words, we do not merely "see" things as they are, but in a context of meaning given to us over many years by our own experiences and those of others.

Thus, while a computer could be taught to "know" that it's cold outside (e.g., via a simple temperature gauge) and then relate this to other things that the computer has been taught, or learned, are cold, only a human can physically experience cold. And, critically, only a person can emotionally relate coldness to other emotionally based experiences. In other words, only humans have a heart; which, interestingly, is precisely where the ancient Egyptians thought logic and rationality were to be found.

Only humans can think about being cold and only humans can link coldness to feelings such as anger, joy, sadness, fear, and surprise that other people have thought about or experienced in the past. Furthermore, no two people will ever experience the same object (e.g., a painting) or event in an identical manner, because the object or event is informed by other objects and events and by our wider value systems and experiences.

Machines do not possess either empathy or morality. Hence, showing a computer a painting such as *Winter Landscape with*

Figures Skating on the Ice will not get much of a response. In the future a computer might be able to recognize the picture and tell you that it was painted in 1643 by Jan Josephsz. van Goyen, but even then, you are unlikely to get an emotional response and even less likely to find that the computer becomes inspired and rushes off to paint something itself.

A good idea?

The brain works through electrochemical processes and thinking occurs within the cortex. The development of neurobiology and the use of functional magnetic resonance imaging technology (fMRI scanners), positive emission tomography, and quantitative electroencephalography gives us more information. We can see the blood flowing to different parts of the brain during different activities and, thus, we are starting to see how brain activity influences behavior. However, we cannot look at an image of the brain and say with any degree of certainty whether conscious or subconscious thinking is taking place.

A recent discovery is that loss of concentration may be linked to changes in brain activity. In certain circumstances a loss of concentration can cause quite serious accidents, so if you can predict when it will occur you can, in theory, take steps to prevent the accident from actually happening. This isn't exactly time travel but it's getting close.

Discoveries like this have got some people very excited because they think they will be able to predict and control the minutiae of human behavior. They believe (and perhaps they are not totally wrong) that there is big money to be made from the intersection of neuroscience and business, although at this stage the technology is not a practical proposition for the likes of IKEA or Coca-Cola. As David Brooks, a cognitive neuroscientist, points out, it's "like flying over Los Angeles at night, looking at the lights in the houses and trying to guess what people are talking about at dinner."

So we still have next to no idea how human creativity happens and we are still a very long way from being able to predict originality or develop a set of generative rules for problem solving or idea creation. Science also has next to nothing to say about our imagination and, when scientists talk about human consciousness, they can barely define it, let alone explain it.

To some people this matters, because they would like to build machines that will one day be capable of creating ideas. Other people would like to understand more about how original thinking happens because there is a hard cash value on inventiveness.

Thinking ahead

Let's return to the issue of AI and consider where developments in this area might go in the future. The term "artificial intelligence" was invented by American computer scientist John McCarthy in 1956. He envisaged an age of intelligent machines that would be reality within a decade. As is usually the case with such predictions, his "what" turned out to be roughly right but his "when" was way off.

We now have basic forms of machine intelligence, although we barely realize it. If you mistype a word in a Google search, say "problem sloving," you will be asked if you mean the nearest equivalent, "problem solving." As another example, you can buy a camera with automatic face recognition. Program a few familiar faces into the camera and it will hunt out these faces in a crowd to ensure that it captures them sharply. Sounds great. But this is yet another example of how digital technology is reducing our field of vision. The more we focus on what is already familiar (through personalization), the less we will notice, or care about, everything else. There is even a toy for 8 year olds that allows players to manipulate an inanimate object using just a headset and human brainwaves—mass-market brain-to-computer telepathy in a toyshop near you for under $300. That is now; imagine what's next.

Scientist and science-fiction writer Victor Vinge wrote an essay in 1993 saying that the acceleration of technology was leading to a development "comparable to the rise of human life on Earth." The idea here is that much in the same way that computer processing speed and memory are accelerating at close to an exponential rate, technologies such as AI, robotics, and nanotechnology are similarly accelerating. This will result, according to observers like Ray Kurzweil, in the arrival of posthuman hybrids and machines so intelligent they will be able to design themselves (known as the technological singularity). According to Kurzweil, who set up the Singularity University with backing from Google, this will probably be by the year 2045. Kurzweil also believes that within his lifetime (i.e., before around 2050) it will be possible to upload the contents of the human brain (and thought processes) onto some kind of machine and have the ensuing "thing" know it's there, thereby creating a kind of digital immortality.

Is it possible to outsource the human brain? Computers are already analyzing data and recommending actions; that is, making decisions. GE has been using decision-making algorithms since the 1990s and firms such as Mu sigma, Market RX, and Infosys are starting to develop credible decision-making software. Questions or decisions that are outsourced include where to build new factories and shops, how to set prices for subsets of customers, or where to spend promotional budgets.

This won't be the only thing on our minds in the future. As well as external machines that think, it is more than likely we will be manipulating our brains from the inside too. According to Alex Soojung-Kim Pang, a research director at the Institute for the Future in California, it is becoming more common for healthy people to take drugs to increase mental productivity.

You may not realize it, but you have almost certainly taken at least one biochemical drug today to improve your mental performance. I had a mug of strong tea at breakfast followed by several cups of coffee, all of which were taken with the specific aim of being more alert and focused. This evening I might have a glass of wine to relax—or to increase my imagination quotient.

A few years ago a product called Children's Smart Oil was ranked #4 on Boots the Chemist's list of top 10 vitamins and supplements and there are currently about 40 cognitive enhancement drugs in development in the US. Most of these aim to cure diseases such as Alzheimer's, but a few are trying to enhance the already well, by increasing learning ability or memory. The US Air Force is interested in drugs that can counter fatigue (it has a unit called the Fatigue Countermeasures Branch) and businesses would undoubtedly embrace new drugs if they were legal and commercially productive.

You might wonder whether it's fair for people to do well at work or school just because they can afford mood- or mind-altering drugs such as Ritalin and Provigil. But many people already benefit financially from surgery that makes them look younger or more attractive, so is there any real difference? It's not only younger people that these drugs appeal to, either. Given the increasing number of people aged 60 plus who are losing their capacity for recollection, the science of memory recovery and preservation is set to be a growth industry of the future. Conversely, the removal of memories of terrorism or war or widespread technologically induced anxiety is also likely to receive an increasing amount of interest and funding.

From a purely evolutionary point of view, memory is used for future survival. Our brains are programmed to remember good and bad experiences so we can repeat or avoid them. Moreover, if an experience is repeated often enough it passes from short- to long-term storage. Thus the idea of directly implanting a specific memory, or general feeling, into the human brain through the use of technology (downloading as opposed to uploading), or through the use of pharmaceuticals, is a prospect that's both lucrative and alarming.

One way of removing unwanted memories is to administer drugs immediately after a dramatic event. Another way is to turn on or off certain genes that are involved in the consolidation of memories. There is clearly a rather fanciful element to the idea of popping a pill to "change your mind," but we are steadily moving in this direction.

One of the most successful drugs aimed at increasing concentration is Modafinil, which was launched in 1998 and now has sales of $500 million a year. It is intended to treat medical conditions like sleep apnoea, but it is increasingly being used as a stimulant, much in the same way as coffee, alcohol, cigarettes, or even cocaine. Not surprisingly, the US military is interested in Modafinil (and newer drugs like CX717) because it can keep combat troops awake for 48 hours. Is this weaponizing the human mind?

Perhaps one day we will see governments downloading imagined experiences into terrorists' brains as punishment, or to induce them to admit to certain crimes. Maybe we will even have a way of monitoring individuals 24/7 to analyze location, purchasing behavior, and thought patterns so as to anticipate future crimes. Governments partly do this already with suspected terrorists, so perhaps one day the idea will be extended to everyone else.

Do we own our own thinking?

While the fundamental nature of human intelligence and consciousness is still largely a mystery, it is already quite clear that our interaction with objects and environments is the bedrock of human intelligence and thinking. Everything we think and do, every single idea or thought we have, is connected in some distant way with things we have historically done, seen, experienced, or thought before. We are, as it were, all carbon-based machines. And the less aware of this we are, the stronger and more influential external factors and our subconscious autopilot become.

So, if manmade machines are likely to develop significantly in the future, surely we should expect our minds to change in some significant ways too? This certainly seems to be the view of Susan Greenfield, who says: "If you buy into the idea that the mind is the personalization of the brain, the organization of neural connections through experience, then that brain will be highly vulnerable to 21st-century technology."

The future of mental enhancement is uncertain and the same is true for mental privacy. Quite frankly, if an adult chooses to do something to their own brain one could argue that this is largely their business. But having other people do things to your brain without your consent or knowledge is something else entirely.

When we think of privacy we usually think of the right to be left alone to do what we like, so long as we do not interfere with the right of others to do the same. In particular, we're referring to the right to do more or less what we like in the privacy of our own homes. But what about inside our own brains?

Clearly, an idea expressed in a physical or virtual form can have some kind of intellectual property rights, but what about thoughts that are simply thoughts? The right to own the thinking that comes into our heads and to decide what stays there and what doesn't is surely a very profound form of privacy or intellectual property.

Currently outsiders cannot penetrate our minds unless we let them in, but this may change. In the future, mental privacy may be as dead as locational privacy and our spending patterns are today. Spam that was once only on email is finding its way onto cellphones and the next step may well be inside your head (I think therefore I spam?).

Your phone company (which already knows where you are because your phone is switched on) could analyze your voice/text conversation and pick up the word "hungry." It could then instantly work out that there are several food companies in close proximity and send messages to all of them asking them to bid on the right to inform you of their presence.

Or perhaps you are strolling along thinking about what you will do that evening when a commercial message is sent straight into your head. Annoying? Certainly. Impossible? Certainly not. The technology already exists to broadcast sound into your ears, whereby only the person walking directly into a narrow beam of sound will hear the message. The message could even be personalized. So what were once relatively quiet, private thinking places will increasingly become noisy and commercialized, either directly or through messages being sent to the various digital devices we carry.

GPS and a host of other remote monitoring technologies such as RFID (radio frequency identification tracking tags) will put increasing pressure on individuals to expose yet more elements of their everyday lives—and their everyday thoughts—to public view. This information could then be sold or exchanged for other information or for money. The issue of who ultimately owns this data (that is, our thinking) is significant. Will it be the individual who generated it, the device that created it, or the governments and corporations with an interest in analyzing and archiving the patterns deep inside the data? Think about that the next time you use Google or send a text.

Using machines to find out exactly what people are thinking at any given moment is a long way off, but using machines simply to observe where people are or what they are doing is already relatively easy. Of course, one can lead to the other. If you know that someone is attending a far-right political rally or is buying books about how to make explosives, you have an insight into the direction in which their thinking is heading.

Let me give you an example. In 2008 the membership list of the extreme-right British National Party (BNP) was leaked online in the UK. Someone then mashed up this information with some mapping data so that people could enter their postal (zip) code and discover which, if any, members of the BNP were living in their immediate vicinity.

What was once very private can now become very public. Data have been stolen and published for centuries. What is totally new is the utter ease and speed with which this can be done and the number of people such information can reach. In cyberspace, everyone can hear you scream.

You could, of course, argue that the circulation of such information is healthy. The transparency of thought brought about by digitization and connectivity is making the world a more honest and ethical place. This may be true, but where do you draw the line between public and private good? Moreover, what happens when incorrect information is circulated? One of the consequences of rapid information transmission is that we increasingly fail to think properly about the validity of incoming or outgoing information; we are too busy and there is too much of it.

Remote "mind reading" may be a way off, but perhaps we can use machines to read minds in another way. Companies are already screening employees for signs of drug taking; what if, in the future, they made employees undergo brain scans to uncover hidden race bias or criminal activity? An organization called Cephos uses fMRI scanning to divine the truth—quite literally. According to its founder, Dr. Steven Laken, lying makes certain parts of the brain light up under scanning. He believes that the US courts are already close to accepting such scans as admissible evidence.

Historically we have been free to think anything we like so long as our actions do not break the law. But what if the act of thinking certain things itself became illegal? Mental privacy will be one of the hottest issues of the twenty-first century and we have only just begun to debate who or what is allowed to look at our brains. To think that in the 2010s one of our main concerns was receiving too much email!

Out of sight but (perhaps) not out of mind

So far so good. I would now like you to cast your mind even farther into the future and suspend your judgment about what is and is not possible. I want to consider the division between mind and body.

Our world can be divided into two: objects (things) and thoughts (minds). The direct implication is that one is tangible and the other isn't. But what if your thoughts were real in the sense that they could float off into the air or become physically disconnected from your body? What if the human mind were capable of being nonlocal to the human brain? What if it were possible for your mind to somehow become unglued and float outside of your body?

This is fringe thinking, although the idea does go some way to explain phenomena such as out-of-body experiences. For example, there have been some experiments suggesting that an unconscious mind can see real things in a room that a conscious mind couldn't possibly see.

It is possible that our brains might be considerably smarter than we currently imagine. Perhaps two minds, separated by a considerable distance, may be able to communicate with each other on some level, such as direct brain-to-brain telepathy. And if you think this is too fanciful, remember that current science has found it is possible for a particle to be in more than one place at the same time.

When it comes to our future minds, things could one day be proven that our current minds struggle to comprehend or believe. Can you imagine Elizabethans trying to get their heads around quantum physics? We find it extremely difficult to think about things that have no reference point to something else we already know.

Few, if any, of the inner processes that are used by our brains are ever made visible to our conscious minds. The vast majority of what we do is done without us consciously thinking about it. Conscious reasoning is therefore pretty much an illusion—or is, more often than not, the process of retrofitting decisions to make us feel good. In other words, we rarely use logic alone to do anything and many of our decisions, including key ones such as whom to marry or which house to buy, are made unconsciously. Our conscious mind is involved only in the sense of trying to postrationalize or justify the decision microseconds, minutes, or maybe months later.

As psychologist Cordelia Fine puts it: "Never forget that your unconscious mind is smarter than you, faster than you and more powerful than you." Our brains distort and deceive. They defend and glorify our egos. Given half a chance, they can also get in the way of deep thinking.

It looks as though our brave new digital world will indeed create a new type of thinking, although the phrase "mental processing" or "power browsing" might be closer to the truth in many instances. This is the type of thin, frenetic thinking where we are aware of what we are doing but we are not thinking deeply about how or why we are doing it. We see but we do not comprehend.

If we are so smart, why do we seem intent on giving our minds away to an array of digital distractions? Why doesn't society have more time for slow thinking and single-tasking? And why are we allowing machines to destroy the very things that make us human and make life worth living?

However, we are not hapless victims of technological change. New technologies often overshoot in terms of ambition and effect and we have a tendency to correct any long-term imbalances when we finally see them. And although this might be true, are you willing to bet your mind on it?

If we are going to be deep thinkers, the place at which thinking needs to begin is in the unconscious mind. But how are we going to tap the subconscious for that deep thinking? The next chapter explores the sex life of ideas.

Chapter 4

The Sex Life of Ideas

"Thoughts, like fleas, jump from man to man. But they don't bite everybody."

Stanislaw Lec

When we consciously try to solve a problem or generate a new idea we are often unsuccessful. We hit a mental block. However, once we have prepared a problem and left it alone, our unconscious mind (the brain's problem factory, if you like) starts work. Ideas begin to fly around in our head, banging into each other and forming novel combinations or mutations.

A quote from William James, the father of modern psychology, sums up the process very well:

"Why do we spend years straining after a certain scientific or practical problem, but all in vain—thought refusing to evolve the solution we desire? And why, some day, walking in the street with our attention miles away from the quest, does the answer saunter into our minds as carelessly as if it had never been called for—suggested, possibly, by the flowers on the bonnet of the lady in front of us, or possibly by nothing at all that we can discover?"

Where do ideas come from, daddy?

Memory plays a key role in the generation of ideas. This is because ideas are rarely new. More often than not new ideas are actually a recombination of old ideas or existing thinking. For new ideas to be born you need two or more old ideas to jump into bed and get a bit frisky.

What is more, ideas can only select their genetic parentage from ideas, and variants of ideas, that already exist. Thus thinking, and idea generation in particular, is associative. One of the key traits of highly creative individuals is the ability to combine unrelated elements to create new concepts. A key driver of this is experience. The physical clustering of like-minded individuals is also important, because if people are close together then ideas can easily jump between them.

Interestingly, this recombination is rarely done consciously. Our brains continually soak up information—every single experience we ever have—and then file it away for future use. Furthermore, these historical experiences do not just gather dust in some remote corner of our brain. They actively influence how we think and act on a day-to-day basis, although most of the time we have no idea that this is happening.

Dr. David Rock and Dr. Jeffrey Schwartz have discussed the relationship between memory and conscious attention at length in their essay "The neuroscience of leadership," of which a quick synopsis is in order. When we encounter something new, information enters the brain's working memory area in the prefrontal cortex. This is an energy-intensive process so, once the brain recognizes something as familiar, it is sent instead to the basal ganglia, a much less energy-intensive space. The basal ganglia deal with familiar or routine activities and information and their existence explains how we can do certain things rapidly, usually without "thinking." However, here's the rub. Because the basal ganglia are so energy efficient, there is a tendency for them to run on autopilot, with habit becoming hardwired into the system.

A good example of this is driving. When you first learn to drive it takes a lot of conscious effort. But once you've mastered it, your subconscious takes over to the point where you can literally drive without really thinking about it—and you often think about other things instead. It is only when you are faced with a new or unexpected situation (or something that's potentially dangerous) that this autopilot switches itself off. Visit a different country and drive on the other side of the road in an unfamiliar city, and you revert to conscious and deliberate thinking once again.

The first stage in generating an idea is often referred to as incubation or fermentation, and generally starts when we stop thinking about the problem at hand or become involved in a totally unrelated activity—or, better still, when we start doing what appears to be nothing at all. This stage can last for a few hours, a few months, or a few years. But eventually a solution or idea suddenly pops into our head, seemingly out of nowhere. Patience is key, because the process can be a bit hit and miss.

If you want a big idea or a clever solution, there is simply no substitute for waiting. And therein lies a major problem. If our lives are becoming busier and we are continually being distracted by a plethora of digital devices, when, exactly, can this waiting take place? If we never switch off our iPhone or BlackBerry, when can we think about nothing for an extended period? And if we cut down on our sleep due to the demands of work, when can we awaken our subconscious?

Slowing down and switching off occasionally is one way of increasing mental productivity, but there are other ways too. One is happiness. This might sound crazy, but the brain is more receptive to new information when we are in a good mood. Research suggests that our mood directly influences how, and what, we think. For example, in one study, awareness of the weather on a particular day (either good or bad) was shown to directly influence people's overall satisfaction with their lives, which, in turn, affected the clarity of their judgment on other matters.

Even smell can do it. In another study, pleasant smells dulled the perception of pain and thereby altered mood. Whether pumping pleasant smells into an office makes people more mentally productive is interesting, although the idea of using manufactured fragrances to manipulate the mood in an office makes me feel rather queasy.

Because our thinking is fundamentally influenced by our emotions and, in turn, our moods, things that influence our mood have a direct impact on decision making and idea generation. One day we might build machines that can judge what mood humans are in and adjust themselves accordingly, but until then, machine and human thinking are quite different. If you put a fresh rose in front of your laptop it doesn't respond.

As an aside, try this devious little experiment the next time you want to sell an idea to your boss or ask for a day off. First of all, pick a sunny day. Instead of getting straight to the point, start the conversation with a remark about how nice the weather is. Then move on to a little light banter about a mutual friend you both like. In theory, this puts your boss in a good mood subconsciously and he or she is more likely to agree with you or accede to whatever it is you want.

Mental impasses and gridlocks

Recent research suggests that early blockages to thinking are sometimes linked with strong gamma rhythms in the parietal cortex, the part of the brain concerned with integrating information. The brain sometimes becomes gridlocked, possibly because there is too much information passing through or, more likely, because excessive attention creates a mental impasse or wall. This links directly to the amount of information we are now being exposed to and the need to continually scan the digital as well as the physical environment for new opportunities and threats.

One solution to these mental gridlocks is straightforward: just stop thinking for a while. The kind of deep thinking you need for creativity or problem solving is directly connected to slow thinking and no thinking.

Today I spent an hour having a Thai foot massage; all in the name of research, obviously. For the first ten minutes nothing much happened and my mind was running at a thousand miles at hour. But then things calmed down and I suddenly remembered something I had forgotten to add to the book!

Relaxing or thinking about something else can help to break a gridlock. You have to consciously shift your attention. Looked at another way, when the brain reaches a roadblock it needs either to wait patiently for the blockage to be cleared or to take a detour in a different direction. But when your brain is speeding fast down the information superhighway, stopping by the side of the road for a quick break is virtually impossible. You'll just get run over.

The relaxation requirement is true for creative logjams and it also works for memory jams. For instance, how often have you forgotten someone's name only to have the answer suddenly pop into your head hours or days later when you were busy not thinking about it?

This process has been known about for many years. Henri Poincaré, a nineteenth-century French mathematician and physicist, talked about hidden combinations of unconscious ideas and described a mental process borne out of personal experience that involved preparation, incubation, illumination, and verification. Here he is on unconscious incubation:

"One evening, contrary to my custom, I drank black coffee and could not sleep. Ideas rose in crowds; I felt them collide until pairs interlocked, so to speak, making a stable combination."

Poincaré's account of the mind's problem-solving abilities
works well in areas of mathematical or scientific creativity where
there are rigorously articulated problems, but what about business decision making where problems are open-ended, or artistic creativity where problems don't exist at all? To what problem
is Shakespeare a solution and can the answer be verified
empirically?

Being unconventional

Chapter 2 talked about the importance of childlike thinking and
having an open and nonjudgmental mind. For instance, people
use story telling to sell new ideas to cynical audiences because we
tend to connect best with new ideas that contain elements of
personal experience. Story telling also includes humor, which
has very strong links to original thinking. This is a point made
by Edward de Bono, who has famously written about how the
use of lateral (as opposed to vertical) thinking can lead to novel
and unexpected solutions.

The reason jokes are generally funny or clever is that they initially send us down a familiar or expected path but then, at the
last minute, turn around or reverse things in a totally unexpected way. Jokes are disruptive. They suspend logic. They suddenly break context and leap laterally to another, totally
unexpected destination or domain.

Original ideas seek to do exactly the same thing. They break
the mold of conventional thinking and end up somewhere
totally new. New contexts create new possibilities. Arthur
Koestler summed this up when he said:

> *"Humor is the only domain of creative activity where a stim-
> ulus on a high level of complexity (a joke) produces a...
> sharply defined response on the level of physiological reflexes."*

The American comedian Steve Wright, US cartoonist Gary Larson, and British cartoonist Matt are masters at breaking context or combining domains.

Interestingly, US scientists have now shown why the funniest jokes are the most difficult to remember: They subvert common thought patterns by using very complicated or unexpected turns. In contrast, less funny jokes are easier to remember because they use more predictable structures that are simpler to follow.

Quantity is quality

When we are born our brain is already eight months old, at which point the external physical environment starts to shape it and every future experience contributes in some small way to the person we will eventually become. At age 6 our brain is already 95 percent of its adult weight and it continues to develop until between the ages of 22 and 27, after which it goes into a slow but terminal decline. While knowledge and expertise tend to increase with age, originality tends to peak after people reach their 30s. So if you are after new ideas (as opposed to wisdom) you should seek out people under 30.

A love of new ideas and experiences isn't confined to young adults, either. A fetus will respond to new experiences (e.g., sounds) but will tend to ignore the same experience after a short while. This is called habituation and the principle carries on throughout our lives. As David Rock and Jeffrey Schwartz put it: "The basal ganglia are running the show." Entrenched neural circuits mean that we are quite literally hardwired to resist new ideas. A further problem is that new information and experiences can trigger our fear circuits, which are located in a part of the brain called the amygdala. In short, we get very scared and our animal instincts take over.

One of the encouraging conclusions I've reached from researching this book is that quantity is quality when it comes to the generation of new ideas or fresh thinking. Most of us are creatures of habit, so it is vital to be exposed to new sources of information and experience from the earliest possible age. To be healthy and to continually come up with new ideas, our minds need constant stimulation. New experiences build up a rich bank of nutrients, which not only influence our mental health but our future thinking and ideas. Curiosity is vital to this. Chance is important too, although noticing something and noticing that you've noticed it are two very different things.

Science writer Margaret Bowden makes this point when she says: "Serendipity is the finding of something valuable without it being specifically sought." Biochemist Albert Szent Gyorgyi put it slightly differently: "Discovery is an accident meeting a prepared mind." You need to walk around slowly with your ears and eyes wide open. Take the time to notice things and to reflect on their implications and consequences.

Surprisingly, formal intelligence can have very little to do with original thinking. For instance, when it comes to invention and entrepreneurship, a pair of well-traveled eyes and a developed sense of wonder will often take you further than any formal education, although it is usually best to have both. Then again, perhaps the real issue is our rigid and rather outdated definition of intelligence.

Convergent thinking is logical thinking. Its objective is to find single, logical, and correct solutions to clearly articulated, logical problems. It tends to be straight-line thinking, which relates back to known answers and solutions. This is the world of clear right and wrong so historically beloved of education and applied science.

However, there is another type of thinking called divergent thinking, which is about finding multiple, novel solutions to new, ambiguous, or poorly articulated problems. This is deep thinking: the world of originality, systems thinking, flexibility, invention, and ideas.

Moreover, while machines are not yet capable of human-style ingenuity, they are becoming very clever indeed at tasks that can be reduced to a formal set of logical rules or historical knowledge. If these trends continue, then logic, analysis, and expertise will be increasingly outsourced to low-cost countries, or outsourced to machines much in the same way that human muscle was back in the nineteenth and twentieth centuries. Thus, the future will belong not to logical men and women who can acquire, digest, and regurgitate data, but to deep thinkers and innovative organizations who can spot anomalies, ask original questions, and dream up fresh ideas.

Furthermore, digitization and global connectivity will distribute knowledge everywhere in the future. Knowledge used to be rare. Several hundred years ago it was expensive to acquire and was therefore a source of power and influence; not any more. Knowledge is becoming instantly available to almost everyone. One skill that people will need to acquire in the future is not spot knowledge or knowing things per se, but knowing how things relate to one another. Similarly, value will lie less in highly specialist technical knowledge, which will fast become obsolete, but in the ability to create and cross-fertilize ideas to create new knowledge. This is the world of multiple types of intelligence and deep thinking.

Two brains are better than one

The idea of convergent and divergent thinking, born in America in the late 1940s, in turn led to the thought that the brain is split into two and that different processes take place in each of the two separate regions. This idea eventually led to a series of famous experiments on patients with split brains (caused by

injury or birth defects) and the split-brain theory, which won
the psychobiologist Roger Sperry a Nobel Prize for medicine in
1981. Split-brain theory says that the left and right hemispheres
of the brain are responsible for different tasks but that each
depends on the other for proper functionality.

The (logical) left side of the brain processes information. It is
fact based, analytic, quantitative, and verbal. It is very good at
making snap decisions or rapidly processing precise bits of
information. This is our quick-thinking, digital technology-
friendly side. In contrast, the right side of the brain is holistic,
intuitive, synthesizing, emotional, nonverbal. It gives us the bigger picture and is our philosophical or deep thinking side. And
it is this side of the brain that I believe is potentially most under
threat from the digital era.

For example, a study by Professor David Nicholas, Head of
Information Studies at University College London, claims that
kids are being remolded by the web. They are inhabiting a new
throwaway world where concentrating on a single thing for an
extended period (e.g., reading a book) is becoming rare. They
skip from one source or activity to another, using one side of the
brain (the quick-thinking left side) to the virtual exclusion of the
other. If this trend continues, not only will books die out in the
future but so too will extended linear argument, because this is
not something that digital mass media can provide.

The two sides of the brain have different ways of organizing
reality. The left side is concerned with details, the right with
abstract connections. The left is more diagnostic, interpretive,
and wants to find order; the right side is more open and experi-
mental. It's rather like newspapers, where the text delivers the
facts while the pictures place the story in some kind of narrative
context.

A good experiment to demonstrate the left/right split is as
follows. Think of a number between one and five and count the
number out using your hands. Did you use your left hand? 70

percent of people do, because that's the side where the brain does math.

Another example is the way people often look upward and away from another person when they're thinking. In 1972, Robert Ornstein, David Galin, and Katherine Kocel at the Langley-Porter Neuropsychiatric Institute in San Francisco found that individuals always look away from the questioner and in a particular direction depending on the nature of the question. If you're asked to count the number of rooms in your house you will usually look left, whereas if you're questioned about how to spell Mississippi you will generally look right.

However, while the right side of the brain is responsible for many of the key building blocks of creativity, the right side still needs the left side. The right brain is in charge of "What if?" while the left-brain is "Head of No." One side is curious, playful, and childlike, the other is the voice of reason and experience.

Nevertheless, talk of two sides is actually highly misleading. The two sides are intimately interconnected and we cannot function well unless both halves are working correctly. Indeed, it might be better to think of two brains rather than one brain split into two. According to Robert Ornstein, each side of the brain is quite capable of doing pretty much anything, it's simply that one side is much better at some things and vice versa. But it is still better for each side of the brain to rub and polish itself against the other.

Whichever way you choose to think about it, the conscious and unconscious parts of our mind are part of the same system. Furthermore, while there are differences between the left and right hemispheres (or, more correctly, the upper right, lower right, lower left, and upper left quartiles) there are also differences between how male and female brains operate. If you're running a team charged with fresh and original thinking, you should have both male and female brains on the team.

Why clever people make dumb mistakes

Is the digital age making us any smarter at not making silly mistakes? Unfortunately, it appears not. Currently the trends of increased demand for our attention plus a mix of complexity and connectivity are against us.

The reasons individuals and institutions make mistakes are fairly simple and are intimately connected with how our brains work. One reason we sometimes screw up is what is known as "sunken cost." We sink time and money into something and therefore continue to back actions or strategies well past the point of logic in a quest to get either our time or our money back. This is similar to another reason, the "endowment effect," which basically says that we behave differently with things that we own (spending other people's money versus ours). Other reasons include egocentric bias (especially combined with alpha male competitive behavior), confirmation bias (finding facts to fit preconceived ideas), overconfidence, expediency, conformity, and distraction.

Here's a classic example of a mistake, cited by Joseph Hallinan in *Why We Make Mistakes*:

"A man walks into a bar. The man's name is Burt Reynolds. Yes, that Burt Reynolds. Except this is early in his career, and nobody knows him yet—including a guy at the end of the bar with huge shoulders. Reynolds sits down two stools away. Suddenly the man starts yelling obscenities at a couple seated at a table nearby. Reynolds tells him to watch his language. That's when the guy with the huge shoulders turns on Reynolds."

Here's how Reynolds recalls the event:

"I remember looking down and planting my right foot on this brass rail for leverage, and then I came around and

*caught him with a tremendous right to the side of the head...
he just flew off the stool and landed on his back in the door-
way, about fifteen feet away. And it was while he was in mid-
air that I saw... that he had no legs."*

How could this have happened? Easy. Reynolds was distracted.
He looked but he didn't see. He had a view of reality that was
influencing what he saw. How we perceive the world influences
what we actually see.

Here's another example. A few years ago, Christopher
Chabris at Harvard University and Daniel Simons at the
University of Illinois came up with a now famous experiment.
They showed volunteers a video of a group of people playing
basketball and asked them to count the number of basketball
passes made by one of the teams. Around 50 percent of the volunteers failed to spot someone dressed in a black gorilla suit who
walked slowly across the basketball court, weaving between the
players, for nine seconds. However, when the volunteers were
asked to view the tape again, this time not counting the passes,
they saw the gorilla easily.

The key point is that our attention is not an unlimited
resource. It is finite and we should therefore be very careful
about what, or whom, we give it to.

Another general reason we make mistakes is because of the
way our brains tag information. Essentially, all information and
experience receives a tag (a reference number, if you like) and is
filed away deep inside our head. However, information and
experience do not sit quietly by themselves waiting to be called
up when they will be most useful. They are connected to various
emotional thoughts that are also stored away. Normally this isn't
a problem, because we use such material to make decisions and
judgments. But occasionally these connections let us down.
Sometimes misleading memories attach themselves to information, resulting in false pattern recognition or understanding. We

think we understand a situation (based on previous experience) when we do not.

Some people may also make mistakes because doing so is in their genes. Research by German scientists working at the Max Planck Institute in Leipzig claims that there is a mutant variant of a gene called A1 that prevents people from modifying their behavior based on previous experiences or mistakes. The theory is that under normal circumstances, useful behavior is rewarded in the brain with a shot of the chemical dopamine, which is picked up by D2 receptors. However, in some people these chemical receptors are missing, with the consequence that dunce-like behavior doesn't get noticed. Time will tell whether the idea of serial stupidity holds water, but it's an intriguing theory.

Conscious thinking works well when we have the time to process and analyze information, but so often nowadays we do not. In this new world of shorter attention spans, faster decision making, and endless digital distraction, perhaps unconscious thinking is a better way to make some decisions. As Malcolm Gladwell points out in *Blink*: "We live in a world that assumes that the quality of a decision is directly related to the time and effort that went into making it." Moreover, too much information isn't just useless, it's harmful. Hence his observation: "Our mind, faced with a life-threatening situation, drastically limits the range and amount of information that we have to deal with."

Life-threatening situations obviously demand rapid action, and quick or shallow thinking works well for relatively trivial decisions. However, deep, rigorous, reflective thinking is a foundation stone for serious creativity, strategic thinking, and innovation, and I firmly believe that this kind of thinking cannot be rushed.

Big new ideas that have genuine value require time and they require that we make mistakes along the way to refine them. Ideas need us to have time for information to enter our brain in the first place, so our mind can subconsciously work on them, and they need a time and a place to pop out. All of which rather neatly sends us to sleep.

Why we need to take our ideas to bed

The phrase "sleep on it" has been around for a while, but it wasn't until 1953 that people realized the brain isn't switched off when we sleep. It is now clear that when we are sleeping, our brains are busy processing the day's information.

More specifically, the brain is taking recent memories, stabilizing them, and filing them away for long-term storage (moving them from our desktop to our hard drive, so to speak). We are processing information all the time, but through sleep we strengthen memories and actively filter them, separating what's immediately useful from what's not.

Mental fireworks can occur during the daytime, when we're occupied with something repetitive or mundane, but it usually takes darkness for them to illuminate our thinking. In my own case the most productive time to light the blue touch paper and retire is when I've got jet lag, when I'm not quite awake but not quite asleep either. Margaret Bowden, quoting Arthur Koestler, echoes this:

> *"The most fertile region seems to be the marshy shore, the borderline between sleep and full awakening—where the matrices of disciplined thought are already operating but have not yet sufficiently hardened to obstruct the dreamlike fluidity of the imagination."*

Recent studies have suggested that it's not just memory stabilization and storage that are occurring when we're sleeping. Some people now believe that the brain is also actively solving problems and creating new ideas. According to research by Bob Stickgold at Harvard Medical School, people who sleep after receiving information are better at recalling themes and patterns. Sleep doesn't only stabilize and strengthen memories, it appears to extract meaning too.

It is fascinating that we learn and create things when we are fast asleep. This also emphasizes the need to switch off external distractions if the brain is to get its job done properly; darkness and silence are therefore crucial.

Critically, research is also starting to suggest that aspects of memory stabilization and learning do not occur as well, and possibly not at all, when people try to survive on less than six hours' sleep a night. Forgo some of your sleep and you miss out on the day's memories and learning. Unfortunately, people (adults and children alike) are sleeping much less than they used to. Most adults survive on between six and seven hours' sleep, well below the recommended level of eight hours, largely due to the distractions of work and our 24/7 lifestyle. Lack of sleep obviously influences physical health and mood, but it also affects memory, reaction time, concentration, and attention.

According to sleep scientists, a third of adults in Britain have trouble sleeping and a quarter of these people are sleep deprived. In the US people slept an average of 9 hours a night in 1900; today it's 6.9 hours. Indeed, if you are feeling tired it's probably because you are kidding yourself about how much sleep you get. A study in the *American Journal of Epidemiology* found that people exaggerate their sleep. On average, participants in the study slept for 6.1 hours but claimed they had slept for 7.5.

This is problematical for all kinds of reasons. A prolonged lack of sleep may cause the brain to stop producing cells in the hippocampus, a region of the brain associated with forming memories. Research by Elizabeth Gould, professor of psychology at Princeton University, has found that lack of sleep, along with stress, old age, isolation, and lack of exercise, can inhibit the formation of new brain cells.

There are general health risks, too. Experiments with rats at the University of Helsinki in Finland indicate that the body treats chronic sleep deprivation as a threat, which can lead to stress-related illness such as heart disease. Danger levels may be as low as 4.2 hours per night for 10 consecutive nights. Meanwhile, behavioral scientists at Duke University in the US have found that a lack of proper sleep is strongly associated with hostility, depression, and anger. Do you get enough sleep? Are you frequently stressed? Do you take enough exercise? Precisely.

But there is good news too. The siesta is reappearing in some organizations, and companies such as Metro Naps and Yelo are selling lunchtime and mid-afternoon naps to stressed-out New Yorkers in special sleep pods. This is another example of "what's new is the past."

Daydream believer

It's not just that we need more sleep to incubate ideas, we need to daydream more too. When we daydream, a specific pattern of brain activity is activated. This is known as the default network and it is in this brain state that the mind starts to make connections between seemingly unrelated information, ideas, or events.

It used to be thought that when people were daydreaming their mind was relatively inactive, but this appears to be false. Letting one's mind go by looking out of a window or performing a task that is so familiar it requires little mental effort can have tremendous real-world benefits in generating new ideas. As Baldassare Castiglione says in the sixteenth-century *Book of the Courtier*: "What I have dreamed in one hour is worth more than you have done in four."

Daydreaming links to meditation, which encourages awareness of a stream of experience without forcing an individual to
concentrate or to react to any single event or piece of information. In this context, the phrase "keeping an open mind" starts
to take on new meaning.

Nevertheless, daydreaming is under threat from the modern
world and especially within the modern corporation. Looking
out of a window for an extended period at work is seen, like
going for a long lunch, as wildly unproductive. Yet strategic daydreaming (or the strategic use of lunch) can result in startling
insights and leaps of creativity.

Music on the brain

Examples of how dreams have led to the creation of musical
masterpieces are legion and include the story, related by neurol-
ogist Oliver Sacks, of Wagner literally dreaming up the introduc-
tion to *Das Rheingold* while in a semi-sleeping or somnolent
state. Ravel, Mozart, Chopin, Brahms, and Stravinsky similarly
stated that melodies came to them in dreams or daydreams.

The fact that music is abstract in nature is relevant to any discussion about ideas. As Sacks said in his book *Musicophelia*, "the
neural processes underlying that which we call creativity have
nothing to do with rationality. That is to say, if we look at how
the brain generates creativity, we will see that it is not a rational
process at all; creativity is not born out of reasoning." Philip
Ball, author of *The Music Instinct*, says that music is "a remark-
able blend of art and science, logic and emotion, physics and
psychology."

How can music influence the generation of new ideas? One
possible explanation is that electrical discharges from neural
networks might resemble music if they were communicated
acoustically. This could explain how music somehow helps to

highlight information or feelings that have otherwise been forgotten, unnoticed, or repressed. In other words, music may influence thinking in a manner that is similar to dreaming. Dreaming and music can both bypass logic, and both can trigger thoughts deep inside the brain stem or connect visual images in the cortex. Incidentally, if you are wondering whether listening to music while doing something else counts as multitasking, the answer is no. According to Dr. Clifford Nass, a professor at Stanford University, the modality of instrumental music doesn't appear to create any negative consequences.

Brain imaging techniques have demonstrated that while some professional musicians' brains are shaped slightly differently from those of nonmusicians, musicality exists in virtually everyone. Moreover, while listening to Bach certainly won't turn you into Mozart, exposure to classical music appears to enhance the mathematical and verbal skills of individuals at quite an early age.

Remember also the mnemonic power of music. Think back to your childhood or, if that's too far away, perhaps your own children. Were you sung to, or did you sing songs, to remember complex ideas?

Your instant reaction is possibly "No." But perhaps you don't think of music in this way. Surely you were taught the alphabet using a rhyme? A, B, C, D, E, F, G… H, I, J, K, LMNOP. Similar tricks are used with older students to help them memorize everything from periodic tables to US Presidents, and the technique has been used across all kinds of cultures.

Music and language almost certainly have common origins and the use of music and pictures to convey information and ideas goes back to the earliest days. This is partly because it is easier for most people to remember information or ideas via association or stories than as facts alone.

It does strike me that music and stories are underutilized in the organizational context. Music in particular has the power to make people alert, but also to awaken deep thinking. Music can trigger feelings (something Nietzsche, a lifelong depressive, wrote of frequently), which in turn produce thoughts and images in our minds—it is these thoughts and images that can become novel solutions and innovative new ideas. Music is therefore directly related to the production of new ideas.

Distributed intelligence

So far this chapter has considered ideas in the context of individuals on their own. Could the digital age result in a shift away from the individual toward the group and the development of distributed intelligence?

The tendency of large groups to be smarter than any single individual (largely a statistical phenomenon) has been known for some time, although it took James Surowiecki's book *The Wisdom of Crowds* to place associated ideas like prediction markets on the corporate radar. Problem solving (and to some extent idea generation) is more productive when more minds are given the problem. For example, if, at the local fête, you ask an individual to guess how many jellybeans are in a jar, they will usually be way off. But if you ask 100 people to guess, the average of all the guesses is usually quite accurate.

If you give an individual several guesses, the combined average is also likely to be far more accurate than their first guess. The theory, still untested at this stage, is that the phenomenon is not purely statistical but is indicative of how the human brain works. Our brains constantly create, test, and reject new ideas and hypotheses about how things work—or could work—and it is through this process of experimentation that we refine or deepen our thinking.

However, while groups are better than individuals at solving complex word and letter puzzles, in other instances less can be more. According to an experiment conducted by scientists at the University of Illinois, physical teams of two perform no better than two individuals. Teams of three do create a significant performance benefit, but if you go beyond three people there is no additional gain.

If what you're after is a prediction about an occurrence, or testing or refinement of an existing idea, then a distributed crowd is a very good way to go about it. It has also been shown that richly connected networks are especially good at solving simple problems. However, large groups are poorer for creating an entirely new idea.

This seems to confirm what I've been thinking for many years: if you are after highly original thinking, a single talented individual (or a small group of talented thinkers with diverse experience) is a better route than a larger group. Crowds will tend to reject any new idea that does not immediately fit with already known ideas and also anything ugly, unusual, or different, although over time this rejection will fade. Blame those basal ganglia again. For example, groups first rejected and then embraced television programs such as *The Mary Tyler Moore Show*, *Only Fools and Horses*, and *Fawlty Towers*, and movies such as *Fantasia*, *Blade Runner*, and *Scarface*.

People (especially groups of experts) generally loathe new ideas. New ideas always threaten vested interests and are often inconvenient in terms of the mental energy required to understand or accommodate them. Furthermore, as economics professor William Baumol points out, education can thwart new ideas because it indoctrinates individuals in the expert (historical) thinking of any given field. You may be aware of Isaac Newton's statement: "If I have seen further it is by standing on the shoulders of giants." Benjamin Jones, a professor at the Kellogg School of Management, interprets this further to demonstrate the difficulty of innovation: "If one is to stand on the shoulders of giants, one must first climb up their backs, and the greater the body of knowledge, the harder this climb becomes."

As the world becomes more connected, for instance via globalization and digitization, there is a tendency toward convergence in what people know and think. We use the same roads so we end up in the same places. For example, only 1 percent of Google searches proceed beyond page one of the results. Hence vast networks can have attitudes and behaviors that are actually more comparable with a single individual than a federation of independent and original minds. At worst, group thinking can give rise to bandwagons, such as the popularity of 3D movies and Pet Rocks. Or take Wikipedia, the crowd-sourced (or user-generated) encyclopedia, which is a fantastic invention but is open to abuse. Because contributions are essentially anonymous, it is impossible to get a feel for the reliability, legitimacy, or self-interest of the contributors creating or amending the entries. So, for instance, the Dutch royal family altered information that was not entirely flattering to them. The wisdom of crowds is not infallible. It is not always diverse either. In 2009, 80 percent of Wikipedians were men; 70 percent were under the age of 30; 65 percent were single; and 85 percent didn't have kids.

Celebrate serendipity

It is fashionable (in these days of political correctness and anti-elitism) to say that everyone can be creative. Some creativity consultants would also have us believe that, if employees would just follow their processes,™ the floodgates of creativity will automatically open. Yet, despite decades of lateral thinking and creativity coaching, very few individuals or institutions have been able to create a repeatable process that demonstrably improves the quantity and quality of commercially useful ideas.

If it is not possible to invent such a repeatable process, then how will computers become creative? It seems to me there is almost no chance that computers will ever write a sonnet to compare with Shakespeare's or produce a painting comparable to Picasso's *Les Demoiselles d'Avignon* within my lifetime, and probably never.

One reason why, at its most extreme, such originality or deep thinking is impossible to predict or artificially create is because it is extremely complex in systems terms. It is hugely reliant on chance and serendipitous experience. Consciously orchestrating such accidents and lucky collisions is much easier said than done.

Most breakthrough thinking comes from left field (but not left brain!) and is, more often than not, stumbled upon rather than deliberately sought. In this sense creative thinking and innovation are very much like evolution. (Interestingly, Darwin said of himself, "I suppose I am a very slow thinker.") Logical and somewhat predictable forces are at work, but random events, mutation, and time are the real driving forces. Deliberate practice by itself—even the 10,000 hours that writer Geoff Colvin claims are necessary to gain real mastery—is no guarantee of greatness, especially if you want deep insight, invention, or discovery.

Genuinely original thinking in any existing discipline is also extremely rare. Perhaps we should stop worrying about the process of making great ideas happen and instead stop people killing them off (or ignoring them) when, against all the odds, they do show up. This can be extremely difficult because big ideas are at first hard to understand, largely because, as I've explained, they do not directly relate to what is already known or experienced.

There are courageous individuals and organizations out there who want to cultivate this type of deep and original thinking, but most don't. Deep thinking is just too disruptive, risky, and messy for most people, so organizations may subconsciously develop immune systems against it. As immunologist PB Medawar says: "The human mind treats a new idea in the same way the body treats a strange protein: it rejects it."

One of the main enemies of deep thinking is thus a rather combustible mixture of fear and inertia. This is a great shame, but it is not always a disaster. What most organizations (and individuals) actually want is to be better at everyday idea generation and problem solving, what I'd call fast (or thin) thinking— lots of small ideas, quickly filtered and implemented. Continuous incremental innovations can be very profitable and digital technology, social networks, and open-plan offices are very good at facilitating such ideas.

Innovation is not a physical department. It is an attitude and, while innovation requires commitment and funds from the top, it is very much a bottom-up activity driven by culture and open networks. It is messy, unpredictable, and will fight any rigid or inflexible process foisted on it.

So individuals and institutions should be very clear indeed about what kind of thinking they require and what form of ideas they are really after. If you're looking for small tweaks, then leverage digital technology for all it's worth. But if you want radical leaps, you need to tap into deep thinking—and to do this you need to think (deeply) about how to leverage the extraordinary qualities and capabilities of the human brain. The next chapter looks at how to find space and time to think.

10 ways to breed ideas

▪ Thanks for the memory—we remember all experiences and then subconsciously link memories together to create new ideas.

▪ Break mental gridlock—we are hardwired to resist new ideas, so we need to rest, wait, suspend judgment, and remove inhibitions.

▪ Humor breaks conventional thinking and stops the brain from being stuck in old habits.

▪ We need a combination of convergent and divergent thinking: the right brain says "what if?" and the left brain is "head of no."

▪ Make mistakes—it might seem that we continue to make silly mistakes, but it is through mistakes that we encounter serendipity and accidental discovery.

▪ Go to bed—we think while asleep, processing emotions, memories, actions, and problems to create ideas.

▪ Daydream—let your mind go and become deliberately unproductive.

▪ Use collective wisdom to solve simple problems, but individual creativity to come up with highly original ideas.

▪ Music opens us up to feelings and memories, and helps to develop the brain at any age for new ideas.

▪ Deliberate practice can make us more adept at any skill, but genius cannot be learned.

Chapter 5
Thinking Spaces

"There is more to life than increasing its speed."
Mahatma Gandhi

In 1968, William Anders, Frank Borman, and James Lovell spent three days traveling to the moon and were the first humans in history to glimpse its far side. On the fourth lunar orbit, on Christmas Eve, the crew of Apollo 8 saw something else that had never been seen before: an earthrise, a fragile blue planet rising, somewhat optimistically, above an inhospitable lunar landscape.

Instinctively recognizing that this was a significant event, Anders grabbed a camera and took some photographs. These pictures effectively started the environmental movement back on earth in the early 1970s.

Many astronauts have now experienced this view of the earth, but its power is undiminished. The state of heightened consciousness that astronauts experience when they look back at our planet from a great distance away is called the "overview effect." Ed Mitchell (a veteran of Apollo 14 and the sixth man to walk on the moon) said, "when we see ourselves in this bigger perspective… a shift takes place in your perception and you start to think quite differently." Out in space there is a lot of space and this can quite literally change your mind.

I have experienced something similar when my head is in the clouds. Looking out of airplane windows at the earth 35,000 feet below, problems can look very small indeed. With distance comes perspective.

Changes in our thinking can also happen walking into a cathedral or looking out to sea at a seemingly endless horizon. Natural spaces and spacious built environments somehow transport our minds somewhere else. Our thinking moves away from day-to-day concerns onto deeper questions. Our bodies are dwarfed and our minds temporarily fly off to places that are more mentally productive. Or perhaps it's that our minds expand to fill the available space. Whatever happens, our thinking is dramatically changed by the physical environment. We only really have new ideas when we stop trying to have them, and nonwork environments are a fundamental part of this process.

Certain objects and activities can induce a similar state. For example, sitting down at a desk and placing a carefully sharpened pencil on a crisp sheet of white paper can create all kinds of thoughts. (This never seems to happen when I am typing on a computer or watching television.)

Where do people do their deepest thinking?

While I was writing this book I conducted some research to understand more about thinking spaces. The question I wanted to ask was simple enough: "Where and when do you do your best thinking?" The research was not intended to be a serious large-scale study, but I decided to go for 1,000 responses (a round number) in order to be able to break the responses down meaningfully by age, sex, location, profession, and so on. Would there be a difference, for instance, between where men think and where women think? Or what about age? Surely there would be a major difference between 20 year olds and 40 year olds?

But then it hit me. In my experience a typical response rate to such mailings is between 1 and 2 percent, so I'd need to send emails to 50,000–100,000 people. Holy guacamole.

Then I had an idea (in the bath). People are busy and asking them to stop for two minutes to write 10 or 20 words about

where and when they do their best thinking, especially to some-
body they don't know, is a lot to ask. Chances are they would be
interrupted by something far more urgent.

But what if my interruption was itself unusual? What if I sent
out handwritten letters? This book is about too much information,
constant partial attention, and the urgent need for unplugging and
digital diets, so this seemed like a good way to put a little theory to
the test. Would people be more likely to respond to a typed letter
than a phone call, a fax, or an email? Could old technology (handwritten letters) possibly trump digital communications?

This was a great idea for about five minutes. Then it dawned
on me that you cannot really outsource handwritten letters. I
was going to have to handwrite all the letters myself.

In the end I sent out a modest 100 handwritten letters (with
handwritten envelopes), 300 typed letters, and 500 emails, and
made 99 phone calls. That's one short of 1,000 contacts, but I
didn't really expect the Queen of England to respond anyway.

I did, however, write to and get an indirect reply from HRH
The Prince of Wales. According to his Assistant Private
Secretary: "It is well known that The Prince of Wales is always
inspired by the garden at Highgrove and by working outdoors,
including laying hedges."

What happened next was fascinating. My response rate on the
emails was a paltry 5 percent. The phone calls were a total waste
of time—5 responses—although the typed letters performed
much better, with a response rate of 38 percent. Nevertheless,
this figure pales into insignificance compared to the handwritten
notes. The response to ink on paper was a staggering 74 percent.

The answer to why this happened is a mixture of economics
and psychology. The emails were presumably lost in a sea of digital dross. They were consigned to the digital dustbin. When
people are in email mode they are focused on urgent issues, and
it may also have looked like I hadn't bothered to make much of
an effort.

Marshall McLuhan remains right. Despite recent technological advances, the medium is still the message. So if you want to interrupt someone who's busy sending and receiving emails, sending them another email is probably not the best way to go.

While the typed letters got a much better response, they still looked as though I'd mail-merged some bought-in names and addresses and shot out a few hundred letters in a matter of minutes.

It's entirely possible that the people who received the handwritten letters just took pity on me, feeling sorry for someone who was clearly so impoverished that he couldn't afford a computer and a printer. But I don't think so. The letters took a great deal of effort and this shone through. It was almost as though people felt obliged to read the letters, most of which found their way directly onto the desks of the intended recipients and were not interrupted by the army of assistants employed to guard against unsolicited emails, junk mail, and phone calls.

The handwritten letters were also slightly different. My original idea was to write identical letters, but because I was writing them slowly, new ideas kept popping into my head about improving or personalizing the content.

I received 218 responses to this initial mailing. Then I decided to post the question on my blog, send it out with my monthly newsletter called Brainmail, and stick it on various websites such as Yahoo! Answers. I got hundreds more replies, mainly from young people (I asked everyone to state their age), and the responses were generally not very well thought out, perhaps an illustration of what Carl Honoré, author of *In Praise of Slow*, refers to as "hurry sickness."

In the end I received 624 responses and the most common answers appear on page 96, ranked 1–10 in descending order of frequency. In some cases I have grouped similar responses (e.g., "outside" and "in the fresh air" were both counted as "outside").

Here are some comments from the research, a few from people you may have heard of:

"Over the decades, I think that my best thinking has occurred when I am visiting a foreign country, have my obligations out of the way, and am sitting in a pleasant spot—in a café, near a lake—with a piece of paper in front of me."
Howard Gardner, Professor of Cognition and Education, Harvard University

"Usually when I am not working, and most often when I am travelling."
Baroness Susan Greenfield, Professor of Pharmacology, University of Oxford

"As a drummer I am generally required to avoid deep thinking of any sort. So it's probably whilst driving on a motorway, or on the start of a transatlantic flight. I think it's to do with some distractions so that the thinking is a little freer… also there's a nice reward element that can be employed. No motorway fry-up, or extra dry martini, before there's an opening line invented."
Nick Mason, musician and founding member of Pink Floyd

"I love doing household chores: loading the dishwasher, scrubbing the floors, scouring the pans; the polishing, the cleaning. All the time I am thinking of ways to improve upon the equipment; what would bring forward the technology."
Sir James Dyson, English inventor and entrepreneur

"I sit down (usually at home, in my study, in my grandmother's Welsh oak chair) with a sharp pencil and a blank notebook and start to draw out the idea, almost graphically. Give it a shape, a name, some dimensions, some examples to bring it to life. Then I talk about it with people. Find its centre and my confidence in it."
Adam Morgan, author of *Eat Big Fish* and *The Pirate Inside*

"Some of my best and most complicated thinking (thinking with numbers attached) happens late at night when it's quiet and I've had a few drinks. In can also happen in pubs when I'm oblivious to everyone and everything around me."
Douglas Slater, political adviser, playwright, & founder of Stonewall UK

"Lying in bed in the dark, with the white noise generator producing a soothing whoosh, I sometimes have a few seconds of modest insight."
Dr. Seth Shostak, Senior Astronomer, SETI Institute

"I've had creative thoughts while walking down the street, in the shower, on the squash court, in the bathroom (of course), while shaving..."
Arthur Miller, Emeritus Professor of History and Philosophy of Science, University College London

"I do my best thinking when my brain is uncluttered by the debris of modern day detail... is my MOT up to date, did I send that email, which recycling bin does this go into, do my socks match, I must spend more time with the kids, how does anyone get round to using their airmiles, will the neighbor's new extension be a nuisance, where is that receipt, did my boss notice that it was me who cut him up on the way into work this morning... Bizarrely my very best thoughts appear when somehow my brain engages on a single challenge seconds from dropping off to sleep."
Michael Dick, Connections Planning Director, Coca-Cola

"37,000 feet and half way into a gin and tonic."
Patrick Smith, President and CEO, Futurebrand

"When walking through town. Something about the right balance of stimulus and meditation that you get when immersed in a big city. Concentration and distraction in balance and oxygen in my lungs."
Steve Bowbrick, Blogs Editor, BBC Audio and Music

"'Not at work' would be our collective response from the Insights team! We deliberately hold all our ideation sessions offsite in an effort to break habits and surroundings. And always try to stagger a session over night or over a weekend so that people have time to absorb and think outside of business hours—either in the shower, driving, taking exercise, walking the dog—allowing the mind to wander and ponder!

Neil Brooks-Jones, Head of Insights & Planning, Nestlé

"I usually do my best thinking on a morning run—the crisper the air, the better... I like big sky, open fields and lots of fresh air—I guess it's an oxygen thing—but decisions or 'cloudy issues' often feel more manageable after a brisk 50-minute run."

Head of Digital, entertainment company

"I do my best thinking in bed—My dreams are often very close to real life, so sometimes I wake up having solved a problem."

Roger Graef, writer, filmmaker and criminologist

"I do my best thinking 'if' I go to church—away from things and where I am forced to stop working and usually drift off and stop listening."

Liz Handy, portrait photographer

"Not sitting in front of a screen—I tend to find that ideas come to me when I'm sitting on the ferry, in the shower, walking between meetings, listening to the radio... I think it's something to do with having mild diversions between you and the problem... I can more definitively answer the question 'when don't I do my best thinking?'—when I'm surrounded by kids who demand my undivided attention."

Wayde Bull, Planning Principal, Principals

"In a hot bath."

Napier Collyns, co-founder, GBN

"On the running machine."
Joe Ferry, Head of Design, Virgin Atlantic Airways

"Without doubt as soon as I wake in the morning. By organising and reviewing the coming day in my head I find that I am more organised and efficient. This half hour of true privacy also allows for important personal things to present their priorities—which is often hard when you are active during the day. Thinking before doing really works."
Managing Director, financial services company

"Often the 'spark' comes when I am not supposed to be thinking... I let my mind stop being boxed in by whatever I was doing before-hand. That's when it gets to work on its own, and that's when it works most laterally—both in terms of what it 'chooses' to decide to mull on and in terms of connections it makes between things."
Charles Constable, former Director of Strategy, Channel 5

"Initially I always have to go through a personal briefing phase with a new issue/problem/challenge—as being aware of all the facts is clearly essential... But after that I don't usually just sit still and think hard... Rather the issue 'simmers' in my mind and ideas occur, say, when I am walking the dog or driving the car."
Tony Craig, General Manager, Strategy, NFU Mutual

"Probably playing Lego."
Matt, aged 8

"Before falling asleep or half an hour after waking up and in the shower."
People Science Research Director, food company

"Outside with my friends."
Nick, aged 10

WHERE AND WHEN DO YOU DO YOUR BEST
THINKING?
1   When I'm alone
2   Last thing at night/in bed
3   In the shower
4   First thing in the morning
5   In the car /driving
6   When I'm reading a book/newspaper/magazine
7   In the bath
8   Outside
9   Anywhere
10 When I'm jogging/running

No major differences in responses could be attributed to the various generations: Generation Y, Generation X, Baby Boomers, and so on. I had thought that younger people (those under 25, for instance) would say "email," "on the phone," "on the computer," and so on, but in fact their responses were almost identical to people in their 40s, 50s, and 60s. Admittedly nobody who said "gardening" was much under 40, but beyond that the responses were remarkably similar. Digital technology was hardly mentioned at all. It was used to keep in touch, spread ideas around, or develop them, but the initial spark always came when people were disconnected.

The only significant variation was by profession. People who could be regarded as fairly original thinkers (artists, designers, musicians, scientists, engineers, inventors) seemed to tap into outdoor environments like mountains, beaches, and other relatively slow, tranquil, and isolated places. In contrast, people in senior managerial positions in large corporations appeared to prefer indoor environments or urban spaces where they would bump into other people.

I was somewhat surprised to find that there weren't many gender differences. Women were more likely to suggest "reading" and men often said "gardening" or "fiddling in the shed," but apart from that I didn't discover any significant variations.

Again, I do not pretend for one moment that this is a scientific study, so please spare me the emails about sample size, geographic bias, or any of that kind of stuff. Clearly there is room for more studies and particularly for one on a larger scale. You can find a fuller list of responses on my blog (yes, even I have one) at http://toptrends.nowandnext.com/?p=916.

10 of my favorite ways to create deep thinking spaces

- You've got to let it out to let it in. The essence of deep thinking is an uncluttered mind, so you need to get rid of things you don't need, mentally and physically.
- Being by moving water seems to work—it dilutes the effects of the digital era.
- Other forms of movement are good too: walking, running, or using various forms of transport (especially trains and planes where you have to surrender control).
- Get a room with a good (long) view. High ceilings and horizons seem to elevate thinking too.
- Spending too much time indoors can shrink your horizons. Get outside from time to time. Gardening is an especially good antidote to screen culture.
- Rest your mind. Ensure that you are getting enough sleep. The period between full sleep and awakening seems especially fertile when it comes to acute analysis.
- If something is important, or you are trying to think very deeply, use paper not a screen.
- Lie down (seat 2A on VS 201 between Hong Kong and London is my favorite).

⊓ Go out to lunch and drink a good bottle of wine.
⊓ Talk to someone, especially someone you don't know. A conversation is a great way to unstick your mind.

Why we don't think at work

The results of my survey were interesting, although they more or less tie in with anecdotal evidence and other research, such as that produced some years ago by Roffey Park Management Institute. In that study, most respondents said that they had their best ideas *outside* the workplace. Generally speaking the location was either a neutral setting with colleagues or close contacts, or a relaxed setting such as a long train ride or a plane trip (because these can be among the few occasions when privacy still prevails, perhaps). Walking the dog, relaxing on a beach, or playing music also featured strongly, as did "in the bath."

A key learning here from the Roffey Park study was the informality of the settings, in which the mind can simply drift. This fits with the viewpoint of Kurt van Ess from Steelcase, an office furniture company, who has claimed that 80 percent of ideas are ignited in informal surroundings. The Roffey Park report also clarifies the vital role of informal conversations with trusted colleagues, contacts, or friends rather than formal meetings.

Nevertheless, in my experience the most important insights often come from what is not said, and my own study was no exception.

Only one person from my entire study said: "The office very early in the morning before anyone else arrives... or in meetings when a colleague is droning on." One other person mentioned brainstorms (possibly at the office, they didn't say) but that was it. In a year of talking to and corresponding with hundreds of people about where they did their best thinking, only one or two mentioned a workplace environment and one spoke of an office that wasn't functioning as an office. Nobody said "at the office, of course" and nobody said "school" either. For example, Charles Morgan, boss of the eponymous sports car company, said simply: "Walking in the Malvern hills." I believe he meant walking alone. At least, I am fairly certain he did not mean that the entire management team of Morgan Cars went along with him, two paces behind, taking notes and making helpful process suggestions.

We spend millions designing offices and hundreds of thousands employing consultants to run idea-generation and strategy sessions, but most people's best thinking is done away from the workplace.

The very reason people have good ideas outside the office is that, as we've seen, the brain needs to be in a relaxed (nonwork mode) to do this. You've got to stop thinking before you can get an idea. I have attended countless brainstorming sessions over the past 25 years and while they can result in a plethora of ideas, these are usually highly superficial. You're far better off going out to lunch with colleagues or friends.

When I was writing this page I hit a wall. I couldn't think. Foolishly—and against my own advice—I kept trying, but nothing came. Then I decided to take a day off. I went motor racing. I sped around a track at silly speeds and suddenly my mind cleared. I achieved an almost Zen-like tranquility. To quote motor racing legend Sir Jackie Stewart: "To go faster you need to go slower."

Natural thinking spaces

If you want to have new ideas you need to go somewhere ideas can find you. When I was young I used to go on holiday to English seaside towns like Southport and Broadstairs. Dotted along the promenades and piers were benches containing

AN INNOCENT QUESTION

I want to tell you what happened to me one morning while I was poring over the results of my little survey. I was sitting on a park bench re-reading and scribbling all over a copy of *The Right Mind* by Robert Ornstein, when someone came up to me and innocently asked: "I'm intrigued by the fact you are writing in a book—you don't see people doing that very often... what are you doing?"

It turned out that this chap was a structural geologist and we ended up having a fabulous conversation about deep thinking. I asked him where he did his best thinking and he said it was usually when he was running. Fly-fishing, rowing, and walking worked for him too.

We then had a discussion about why so many people said they think deeply while driving, and he explained that driving was to do with spatial geometry. It was all about distances and speed and therefore used a particular part of the brain (his father, it turned out, had been a psychologist), thus freeing up other parts of the brain to think about different things.

His chosen thinking pursuits tended to be repetitious, even tedious at times, and again these activities used very particular parts of his brain. We got to talking about thinking in organizations and he said he believed there were hardly any deep thinkers left in large companies. They had been rooted out by human resource departments because deep thinkers were too disruptive. It was like the quote by TE Lawrence: "Those who dream by night in the dusty recesses of their minds wake in the day to find that it was vanity: but the dreamers of the day are dangerous men, for they may act on their dreams with open eyes, to make it possible."

In his view large organizations prefer night dreamers, which is why most real thinkers end up working alone or as entrepreneurs inside small start-ups. Corporations just can't handle people who think differently.

pensioners looking silently out to sea. They were there in all weathers. But even on cloudless days I couldn't see what they were looking at. Then one day it hit me. I suddenly realized that they were looking at their own mortality in the distance and it was a clear and uninterrupted view.

Oceans suggest infinity. Similarly, piers are about dreaming and reflection. They suggest escape to somewhere else, which, in my experience, is precisely where you need to be if you are trying to solve a scientific problem, have a new business idea, or think acutely about your own future. Being fixed to the land at one end but to nothing whatsoever at the other, piers are also a physical bridge between the real world of things and the imaginary world of thoughts. Looking from the far end of a long pier back to the distant land, we are temporarily disoriented; another example of the overview effect.

Unfortunately, in our fast-moving, high-pressure, high-octane, get-it-done-yesterday world we are losing slow, contemplative spaces like these at an alarming rate. Piers, like pure thought, do not appear to have an immediate dollar value.

Expansive ocean views don't work for everyone, nevertheless. Writing the foreword to John Steinbeck's *Cannery Row*, Elaine Steinbeck mentions that her husband used to write in a small cottage on Nantucket Island, overlooking the Atlantic Ocean. His workroom, however, looked out in the opposite direction over the moors, because the sea, he said, "would lure me." The beach and the sun were, according to his wife, simply too distracting. Steinbeck's other prerequisites for writing were peace and quiet and privacy. He would often shout out of the window to get the children outside to be quiet and eventually built himself a six-sided writing house on Bluff Point (six sides to every story?).

Gardens are another natural thinking space. Like piers, gardens seem to attract people later in life. One reason, again, is perspective: As we grow older our perspectives lengthen and the

GARDENING AS A METAPHOR FOR BUSINESS

Most metaphors about business are about sport or war. This is useful up to a point, but the fatal flaw in these analogies is that both sport and war generally have an end. Moreover, the objective of both is to defeat a clearly defined enemy. Aims and outcomes are fairly rigid. But real life isn't like that and neither is gardening.

Gardening has no end. There is no finish line. It is about a journey, not a specific destination. Moreover, while business and gardening certainly have enemies, focusing on those opponents too much can divert your attention from thinking about the bigger picture.

If you start to think of ideas as plants, your mindset shifts. In this metaphor you plant an idea in a prepared patch of soil that is set within an overall scheme. You water it and watch it grow. But, as any gardener knows, half of your ideas won't. There's an early American saying about gardening that can be applied to ideas: "One for the blackbird, one for the crow, one for the cutworm, and one to grow."

Business, like gardening, is about flexibility and persistence in the face of changing external circumstances that cannot be wholly controlled. If you want a fertile imagination, you need to sow a lot of different seeds. However, even tenacity doesn't always work. Sometimes plants don't grow because they have been put in the wrong place or because pests have destroyed them. Either way, you have to nurse them back to health or yank them out and start all over again. Gardeners are always yanking things out and starting again. So are entrepreneurs and inventors.

Planting things in the right place is also vital. According to McKinsey: "In sectors such as banking, telecommunications and technology, almost two-thirds of the organic growth of listed Western companies can be attributed to being in the right markets and geographies." In other words, a good idea in the wrong place can struggle, whereas an average idea in a perfect spot is likely to do well.

Then there's the opposite problem. Sometimes things grow so fast that they overshadow what's next to them and they have to be moved if both plants are to flourish. The parallel here is with skunkworks, where teams are moved away from the shadow of a parent company. For example, the telecommunications firm Vodafone was created by accident as a tiny division of Racal Electronics. Someone was given the green light to plant something and see whether it would grow. It did, although I wonder whether this rapid growth would have been achieved if it had been left in the shadow of Racal.

Sometimes things grow so well that over many years the soil becomes exhausted and the only solution is to start again. This is not a bad thing, it is just part of a natural cycle. Fields must be allowed to lay fallow every so often if they are to regain their natural health and vitality. This applies to organizations, and it also applies to people. Sometimes our heads become so log-jammed that we can't think. We can't see the dead wood for the overgrown trees.

Radical ideas are like weeds. They grow where they're not supposed to and cannot be cultivated like orchids in a greenhouse. You cannot sow weeds in any meaningful sense; you can only provide the conditions necessary for them to grow, which in most instances means leaving them alone. Weeds thrive on neglect.

If you want new ideas (new thinking) in your organization, you therefore need to recognize what a weed looks like and allow some to carry on growing even when they're in the wrong place.

And if you want creative thinking in your organization, you need to attract and retain overtly creative people. This means that you need to attract at least a few individuals who wouldn't usually be seen dead working inside a large company, and you need to give them some room to be a little different.

whole becomes visible. As the writer Germaine Greer says: "Gardening is all in the future and the less future people have, the more they tend to think about it."

Private gardens are not generally expansive nowadays, but they are usually quiet spaces where one's mind can freely wander. But to do so one must first surrender control to the rhythms and whims of nature. Gardens are places to sow new thoughts. There is always something to be done in a garden, and some of the best ideas come about when our minds are free to get lost.

You don't have to move to the countryside to benefit from a little green space. A 1990s study of a housing project in Chicago, for instance, found that views of greenery not only improved attention spans but lowered levels of domestic violence too. Numerous other studies have shown positive effects on mental health and faster recovery from surgery.

Another example of the power of natural environments on thinking is a study by Marc Berman, a psychologist at the University of Michigan. Researchers sent one group of students into a busy city and sent another group to walk around a wooded park. When the participants returned they were given a series of psychological tests, which showed that the city not only engendered a worse mood but also a lower attention span and a diminished working memory.

The reason is that a crowded city street requires significant cognitive effort. There are cars and pedestrians to avoid and other potential threats. In a park there is less to think about so, in theory, you can think more about other things.

Deliberately designed work spaces

So how do we build workplace environments that foster deep thinking? If, by some serendipitous circumstance, you find yourself in Bern (Switzerland) with a few hours to spare, I recommend a trip to the Einstein House, number 49 Kramgasse. Go inside and you will eventually find a wooden desk next to a small bay window overlooking the street outside. To one side of this window there is a small plaque with an inscription that reads: "Through this window the theory of relativity was born." Is this a throwaway comment from the museum curator, or is someone making a serious point about the relationship between windows and thinking? Windows allow us to transport our minds to mentally productive places, even when the rest of our bodies are unable to leave the room.

Many years after leaving Switzerland, Albert Einstein himself put up a sign on the wall of his office at Princeton University: "Not everything that counts can be counted, and not everything that can be counted counts." In my view, windows count and so too do other reflective spaces. Einstein was a big fan of visualizing solutions and he also saw the value of furious curiosity.

Fast forward to England in the late 1980s, where some more signs had been put up. This time the signs had nothing to do with Einstein. They were hanging, from time to time, on the doors of the managers of a large chain of British supermarkets. The signs read simply: "Quiet please, I'm thinking."

These days we tend to sit in noisy open-plan offices (no doors, no hierarchy, so the theory goes). Most of the time this doesn't matter. Indeed, studies have suggested that private offices, and especially senior people stuck inside private offices in the corners of buildings, are exactly what you *don't* design if your aim is to cultivate innovative thinking.

For example, writer Malcolm Gladwell points to a study that found something now known as the 60-foot radius rule. The idea here is that innovation is inherently social and is best propagated in busy communal areas or where people are seated close to each other. Beyond 60 feet, people tend not to talk to each other, with the result that ideas aren't passed around. Another study by Thomas Allen at MIT found that communication is 400 percent more likely when people sit 6 feet away versus 60 feet. Add another 15 or 20 feet and communication is almost zero.

A good example of designing a workspace to increase the flow of ideas is a BMW factory in Germany. A decade or so ago BMW built an R&D facility on a new site close to the company's Munich HQ. The ideas factory used the principles of serendipity and proximity, reasoning that if someone is too far away from someone they might need to talk to, chances are they won't bother. All of the design offices were built on the same floor, as was the model workshop. This ensured that BMW's designers communicated by the shortest physical route. The result, according to BMW, was that 24 months were shaved off the development cycle for a new vehicle.

Architecture is clearly one way to achieve idea flow through an organization. Designing kitchens, bathrooms, or staircases with the deliberate aim of making people bump into each other is a very practical way to increase the innovation quotient.

Here's what Ed Catmull, President of Pixar and Disney Animation Studios, says about the design of Pixar headquarters:

> *"Most buildings are designed for some technical purpose, but ours is structured to maximize inadvertent encounters. At the center is a large atrium, which contains the cafeteria, meeting rooms, bathrooms and mailboxes. As a result everyone has strong reasons to go there repeatedly during the course of a workday."*

Staircases are an area where serendipitous conversation can take place. Unilever in the UK once spent a considerable amount of money putting artworks into stairwells to encourage employees to take the staircase route, hopefully colliding with someone else along the way. Kitchens are even better places to cook up new concepts.

"Floor casting" is another way of getting ideas to circulate. This refers to mapping people to identify who the creators of new thinking are within an organization and ensuring that these important people are seated in the middle of a floor, or close to a busy communal area, where people will informally and casually bump into them and spread their ideas. This can be done by tracing where ideas generate and develop, or by tracking people through electronic interactions and invisible social networks.

Or you could just play musical chairs. I was briefly a non-executive director of a company based in the north of England. They had the simple but effective idea of moving members of staff around every few months to increase the amount of interaction. As a result, the managing director might be sitting next to the finance director one month and next to the intern the next.

Having people visible and close together can clearly pay dividends in some instances. But I am still concerned that, even when the architecture is open, individuals are walling themselves off from each other through their use of digital technology. Furthermore, while a mosh pit approach can work well for, say, highly competitive sales people, it may not be right for computer programmers who need to think deeply.

No accident, then, that at Google HQ employees are given offices running off a central Main Street. Open-plan spaces allow staff to bounce ideas off each other, but there are also areas for privacy and seclusion when needed.

I was talking to someone from Google whom I met accidentally at a conference in Dubai. He said that Google had plotted the location of every desk in the Google empire and cross-referenced individual desk location with desk productivity. And guess where the most productive desks were? They were exactly where Gladwell said they would be: in the middle of each floor, because this is where the most human traffic occurs. I'm not sure if it's possible to track the productivity of a kettle or a water cooler, but I'd imagine it would be pretty high if you could.

There is no simple answer to the issue of what the ideal floor layout looks like, but there appears to be a slowly building consensus that points toward the old-fashioned office.

Computer scientist Saul Greenberg looked at the effect of randomly allocating individual offices to people who had previously worked in an open-plan environment. One key finding was that people regarded an enclosed office as being of higher status and their productivity and performance tended to increase to justify this reward.

Other researchers have discovered that open-plan offices make people work more slowly and they subconsciously mimic the behavior of other workers in their field of vision. Open-plan offices have also been found to increase stress levels, as well as creating conflict and insecurity due to noise levels, which also increase blood pressure.

One novel solution to this is a chair designed by a couple of renowned French designers. The "closed-plan" chair features wraparound sides that insulate the user from audio and visual distractions. It looks rather strange (a cross between a bad airline seat and an upturned baby carriage), but its maker claims that it will facilitate better thinking in open-plan environments. Do they use the chairs themselves, I wonder?

What we certainly need is a better balance. Open plan works on some levels, but people should either be given the choice of public versus private or they should be allowed to escape the communal when specific tasks demand it.

Nevertheless, what if open plan—with all the distractions that the digital age provides—is all that's on offer? Anecdotal evidence suggests that what staff enjoy the most when they don't have a room of their own is a little bit of themselves. A desk with a small pin board or a bit of wall that they can personalize is a good second best to a personal office or a communal office with private nooks. This relates to comfort more than ideas but, as we've seen, the two are closely linked.

However, even this modest request is unacceptable to some employers these days. A few years ago Australia Post dismissed a call-center employee because the worker had three family photos on the desk rather than the prescribed two. This is precisely the kind of policy that makes honest employees dishonest and drives good people into the arms of organizations that are more sensitive to their emotional needs. And Australia Post is not alone. Modern society loathes people with minds of their own. A messy desk is right up there with looking out of windows or falling asleep at your desk in terms of high crimes against the corporation.

Perhaps "empty space" architects following in the footsteps of Le Corbusier, Gropius, and Pawson should be tried for human rights abuses, as should any designer or manager who makes no room for people's own things. As for organizations such as the US-based National Association of Professional Organizers (seriously), members should be forced to spend one month living alone in a self-storage unit in Boise, Idaho, until they repent their sins. People, as Virginia Woolf reminds us, are most comfortable when they have a room of their own. Where this isn't practically possible they should at least be given a few personal inches or be left alone to create a sense of themselves on their desks—especially when everything else beyond the desk is uncertain and outside their control.

I think therefore iPod

The relationship between sound and thinking is something we should definitely care more about. If you think about your daily routine, most of what you hear is familiar. If you are in an urban environment you will hear a continuous low rumble of cars, airplanes, buses, and human chatter, interrupted by the odd cellphone, dustbin lid, or alarm.

Before the electronic revolution sounds were more natural, more low key, and more discontinuous. These days excessive noise is responsible for 3 percent of coronary heart disease worldwide, according to the World Health Organization. In the UK, this translates into around 3,000 deaths each year due to chronic exposure to noise. As an *Economist* review of *A Book of Silence* by Sara Maitland comments:

> *"the modern world has little room for serious silence seekers... blasting radios, mobile phones, traffic and aeroplanes can be the source of despair."*

Peace and quiet are useful in terms of deep thinking. Proper darkness and real silence awaken our subconscious. If you doubt this, go out into a desert at night. The darkness is confronting. The silence is deafening. Could TE Lawrence have written *The Seven Pillars of Wisdom* in a wet car park in Croydon? I doubt it. This is another example of how our physical environment shifts our mindset and brings deep thoughts to the surface.

Open-plan offices are noisy, and some people deal with this by listening to music on a device such as an iPod. A recent UK study of 120 companies found that 22 percent of employees used their iPods at work, while US studies have found that 22 percent of people listen to music when working. Removing background noise, while signaling to people that you are not to be disturbed, isn't necessarily such a bad thing. The argument against allowing iPods in a workplace—or other environments like schools—is that they are distracting and socially isolating. Individuals are cut off from each other. But one feature of modern society is noise, so why shouldn't we allow people to cut out some of the auditory distractions with noise of their own choosing?

Maybe organizations should invest in quiet rooms or thinking kiosks? Yamaha in Japan has invented a self-assembly soundproof room called the Avitecs Myroom II. It offers individuals a quiet space for thinking, reading books, listening to music, or for making private phone calls. The box measures 4 ft 6 × 6 ft × 6 ft 9 and fits comfortably inside most offices and homes.

A quiet room is one thing, but could one design the sound of an entire city? This thought has occurred to Murray Schafer, the author of *The Book of Noise*. Perhaps a more practical idea would be to spend more time thinking about the soundscape of individual buildings, particularly those that are designed to encourage thinking. Universities, schools, and public libraries would be prime candidates for auditory makeovers, although because thinking spaces are so idiosyncratic, you would have to build a mixture of rooms including soundproofed cellars, workstations with white noise, and cubicles with power sockets for iPods.

Usually organizational culture focuses on vision, but perhaps in the future it will also mean sound and even smell. Companies such as Singapore Airlines, British Airways, and Bentley Motor Cars already use smell to attract customers, so why not use smell to attract employees or to influence their thinking and ideas?

The use of smell at work might even make employees feel better about being at work, as we saw in Chapter 4. Research from Indiana University in America and the Hebrew University in Jerusalem has shown that burning frankincense lowers anxiety and depression. The resin in question contains a compound called incensole acetate and the burning of this stimulates nerve circuits in the brain. There is even a study by Michael Schredl and Boris Struck that has demonstrated that certain olfactory stimuli can result in either dreamy dreams or nasty nightmares. It gives a whole new meaning to the idea of cheese dreams.

Cathedrals for ideas

There is an old story about an American advertising agency called Chiat Day that is worth retelling. One day in 1993 Jay Chiat (the boss) met Frank Gehry (the architect) and together they had an epiphany: If people were working in cubicles inside offices, then surely their minds were similarly trapped inside boxes? The answer, clearly, was to set people and their thinking free.

Doubtless this was a good idea at the time, but the results were far from satisfying. Employees had to work in offices without desks, filing cabinets, or—you've guessed it—paper. Staff even had to borrow communal computers from a central store unit, which had an opening shaped like a pair of giant red lips. Then there was the revolutionary idea that the offices that had previously been occupied by individual staff members should be turned over to their clients' brands. Hence there was a Nissan room, an Absolut Vodka room, but no room for any staff or their stuff.

The environment looked superb, but it somehow lacked substance. It was like an interactive kindergarten for grown-ups. Indeed, the only thing that was missing was a sandpit. Unfortunately the building had no sense of place. It was self-conscious, sterile, and lacked authenticity and integrity. It was also an environment that suffered from overstimulation. There were very few quiet spaces or private areas and this resulted in people taking over some rooms by force. Others worked out of the boots of their cars in the garage. There was indeed a revolution, but not quite the one anticipated by Chiat and Gehry.

In a sense, though, Jay and Frank were genuinely ahead of their time. In a way this was a brave experiment that just took things a bit too far. For example, a study by the Buffalo Organization for Social and Technological Innovation (BOSTI) in the US identified that the physical design of office spaces does have a significant influence on job satisfaction, productivity, and, ultimately, profitability. Jay Chiat had the right idea but executed it in the wrong fashion.

The Chiat Day offices were an extreme early example of removing physical barriers in order to influence thinking, but the idea persists. These days we inhabit offices with whacky brainstorming rooms that contain giant beanbags, oversized felt pens, and brightly colored walls. There is even a new phrase in the lexicon: "right brain meetings." The idea is that colorful furnishings, together with a scattering of juvenile toys and games, will light up the right side of the brain and bring out the original thinking in even the most cynical and battle-hardened accountant. And unfortunately the trend is taking off. Hotel chains such as W Hotels have offered corporate customers "interesting" meeting rooms featuring Etch-A-Sketch toys, puzzles, and even aromatherapy candles.

A plethora of pseudo scientists and consultants have persuaded us that the only way to think laterally is to behave like a child. There is something in this, much in the same way that architecture influences thinking, but we have taken the idea too far. An infantile culture has grown up and adults are expected to behave like kids hanging out in the local mall. What were once quiet spaces have increasingly been torn down or knocked through to create psychedelic rooms and meeting "yurts."

Isn't the digital era overstimulating and distracting us enough already? The world needs calmness and serenity, not more frantic noise and distraction. In my view, if you are trying to generate breakthrough thinking what you need is more process and discipline, not less. The more you constrain your thinking the better the result, both in terms of quality and quantity.

Some time after building the Chiat Day offices in Los Angeles, Frank Gehry was hired to build the new Beckman Center for Molecular and Genetic Medicine at Stanford University. This is a building built on the premise (perhaps the promise) that the architecture of a laboratory can directly influence scientific creativity. It is literally a cathedral for the information age. Gehry has also been responsible for the Strata Center for Computer, Information, and Intelligence Sciences at MIT. To the people putting up the money for these buildings they are multimillion-dollar bets on the influence of architecture on human thought.

The notion, once again, is that mixing up the interiors will blend brains, which will, in turn, create new forms of thinking. Both of these buildings look a little crazy. Indeed, one of them looks like it has arrived from outer space and has been assembled upside down by aliens. In each case the architectural brief was to create spaces where new conversations and connections would occur naturally. Interestingly, both buildings are very light. As Gehry says: "If you remove the light you dwarf the mice." I think he is referring to the scientific staff, although you never know.

Organized chaos

Rewind your thinking to the signs in the supermarket chain mentioned earlier. A giant American retailer now owns this chain and the signs are no longer there. Neither are the doors to which the signs were once attached. Instead, the company (like General Motors, UPS, and many others) operates a clean desk policy, whereby anything left on a desk after 6 pm is thrown in the bin. This may have something to do with security but it's more likely that managers consider that a tidy desk is a tidy mind. Rubbish.

In 2007 the UK government forked out over £7 million on some comedic consultants to devise a program to increase the tidiness of civil servants' desks. The initiative involved the use of black tape to mark out where, precisely, employees (who were presumably used to red tape) at HM Customs and Excise could place certain items. Believe it not, staff were actually given support and advice about where to put things and were asked whether the fruit on their desks was "active" or "inactive," meaning meant for immediate use or long-term storage.

No doubt this kind of thinking was spurred on by books such as *Winning the Fight between You and Your Desk* by Jeffrey Mayer or by the US-based Productivity Institute, which claims (in all seriousness) that a messy or untidy desk is one of the top five management mistakes. Personally, I think reading such books or hiring management consultants that do would be a bigger mistake.

Fortunately, not all consultants are waging war with mess. Management guru Tom Peters thinks that any employee who has to put up with such nonsense is mad not to look for another job. Several sober-suited academics seem to agree with him. In *A Perfect Mess: The Hidden Benefits of Disorder*, Eric Abrahamson and David Freedman (an academic and a writer respectively) argue that a little disorder can be healthy:

> *"Moderately disorganized people, institutions and systems frequently turn out to be more resilient, more creative and in general more effective than highly organized ones."*

These authors confirm that the focal point of disorder is usually the desk and tell the wonderful story of Leon Heppel, a researcher at the US National Institute of Health. Mr. Heppel's desk was, we learn, fantastically messy. So messy, in fact, that he had the habit of every so often putting a sheet of brown paper over the mess in order to create a second layer. Multistory mess.

One day Mr. Heppel was flipping through some papers on the lower and upper levels of his desk and stumbled on letters from two totally unconnected researchers. He suddenly spotted a connection and put one in touch with the other. This eventually led to a Nobel Prize. Had the letters been in a conventional filing system, chances are the connection would not have materialized.

The way people's desks are laid out reflects how their brains are laid out and it would be a mistake to tell people how to organize either. Messiness acknowledges randomness. Moreover, the anti-clutter movement's obsession with tidiness and order (even our storage boxes have to be ordered these days) flies in the face of reality. Real life is messy and uncertain and any attempt to deny this is not only futile and unrealistic, but misses the unforeseen benefits of disorder.

Hence piles of paper, and other flotsam and jetsam on a person's desk, are actually quite an effective form of filing, especially if material is allowed over time to migrate from one area to another. It looks like chaos from the outside, but it's organized chaos to the person who made it and is responsible for maintaining it. Serendipity can also come into play here in ways that would be much more difficult in a virtual office or where someone is working from home. An idea or image stuck on a Post-it® note for no particular reason could spark an idea in the head of someone accidentally wandering by.

Alexander Fleming may have been well known as a talented researcher, but he was a careless and untidy technician. The story of the accidental discovery of penicillin in 1928 and the tens, if not hundreds, of millions of lives saved through its use are well known. But it is interesting to speculate what might have happened if Fleming had been a neatness freak or had attended a two-day residential course on the seven habits of highly effective scientists.

This is not the same as saying that chaos rules. In some instances (a factory production line or operating theater) strict rules and order are necessary. The trick is to find a good balance between chaos and order and to clean up your desk every so often. Let's not forget the words of Albert Einstein: "If a cluttered desk is a sign of a cluttered mind, of what, then, is an empty desk?"

To my mind clutter and creativity go together. Clutter represents individuality and freedom (and happy employees) and is absolutely essential at some level for creative thinking. Interestingly, according to a survey by Ajilon Professional Staffing in the US, employees with cluttered desks earn higher annual salaries than people of the clean desk persuasion.

Paperless offices

One of the biggest banks in Australia is embarking on a "Future of Banking" initiative that has, at its core, the idea of no permanent desks and, of course, no paper. I suppose it's only a matter of time until they get rid of the people too.

Historically paper has always been an important part of office life, but the idea of a paperless office has been a symbol for modernity and efficiency since the early 1960s. The early theory was that computerization would eventually render physical paper in physical offices obsolete. Unfortunately, what happened was the exact opposite. From about 1990 to 2001 paper consumption increased, not least because people had more material to print and because printing was more convenient and cheaper. But since 2001 paper use has started to fall.

The reason is partly sociological. Generation Y, those born at roughly the same time as the personal computer, have started working in offices and they are comfortable reading things on screens and storing or retrieving information digitally. Moreover, digital information can be tagged, searched, and stored in more than one place, so Gen Y are fully aware of the advantages of digital paper and digital filing. All well and good, you might think—but I'm not so sure.

One of the great advantages of paper over pixels is that paper provides greater sensory stimulus. Some studies have suggested that a lack of sensory stimulation not only leads to increased stress, but that memory and thinking are also adversely affected.

For example, one study found that after two days of complete isolation, the memory capacity of volunteers had declined by 36 percent. More worryingly, all of the subjects became more suggestible. This was a fairly extreme study, but a similar principle could apply to physical offices versus virtual offices or information held on paper versus information held on computer. Digital files or interactive screens could actually reduce the amount of interaction with ideas. I'm not suggesting that digital information can't sometimes be stimulating, but I am saying that physical information (especially paper files, books, newspapers, and so on) is easier on the eye.

Physical paper is faster to scan and easier to annotate. Paper also seems to stimulate thinking in a way that pixels do not. Indeed, in my experience the only real advantages of digital files over physical files are cost and the fact that they are easier to distribute.

There are some forms of information that do need to be widely circulated, but with most the wider the circulation list, the lower the importance of the information or the lower the real need for action or input. As for the ability to distribute information easily, this can seriously backfire. Technology is creating social isolation because there is no longer any physical need to visit other people in person. Offices aren't only about work, any more than schools are only about exams. Physical interaction is a basic human need and we will pay a very high price if we reduce all relationships (and information) to the lowest-cost formats.

This reminds me of a friend who went to Japan a few years ago. One night he went to see a film and to his amazement found someone using an abacus to count the number of people entering the cinema. He asked the man why he wasn't using modern technology and the man replied: "It would be in nobody's interest."

Moreover, when organizations have tried to be efficient by computerizing their records they have often found that paper is actually more effective than pixels.

For instance, the British Home Office once instigated a project that gave police constables on the beat laptop computers. Logically, this was a very good idea. But unfortunately, the project backfired because police officers preferred to look people in the eye when they were talking to them rather than glancing up

YOU DON'T NEED TO WORK HERE TO WORK HERE

Best Buy is a well-known company in the US, Canada, Mexico, and China. It sells consumer electronics and has been rated "Company of the Year" and "Speciality Retailer of the Decade," and named one of "America's Most Generous Companies."

On the face of it Best Buy looks like just another well-run Fortune 500 company, but there is one thing that marks it out as different: you don't need to be there to be at work. Most companies don't trust their employees and because of this their physical presence is generally mandatory. In some instances this is necessary, but in others it is preventing people from coming up with new ideas.

In most organizations going for a long walk in the middle of the morning is frowned on, to put it mildly. So too is going to see a movie in the afternoon or deciding to leave at 2 pm without notice because your brain has frozen up. But none of this is the case at Best Buy. The company has developed a clockless culture where people are rewarded not for the hours they put in but for the work they produce. And if being productive means late mornings or long lunches, so be it. Productivity in departments where the initiative has been implemented has risen by as much as 35 percent and staff turnover has declined dramatically.

Since technology allows people to walk around with their office in their pocket, surely they should be free to work whenever and wherever they wish. But this isn't the whole story. What this particular idea is about is the principle that ideas can happen anywhere and that inspiration for ideas is generally external to work.

Of course, there are downsides to clockless cultures. Teamwork and camaraderie at Best Buy are apparently affected, as is the ability to physically bounce around ideas. Nevertheless, according to one insider, "What we get back far outweighs anything we've lost." Time will tell whether the physical removal of people from a central location affects their loyalty or ultimately their thinking.

while they were typing. Technology, in this case, was getting in the way of truth and clear communication.

A similar example comes from a maker of telecoms equipment in the UK. The company thought it would be a very efficient idea to computerize its salespeople's records. But it turned out that the material salespeople found most useful in making a sale was not something they wanted to put on a database because it contained highly personal information.

The moral here is that organizations tend to measure the things that are easiest to measure and to automate what's easiest to automate. By computerizing or digitizing information we are at risk of hindering the very things that we should be championing. For example, personal relationships tend to deteriorate if they are at the mercy of automation or outsourcing.

I had an accountant once who used to send me a handwritten birthday card every year. One year he sent me an electronic card (aided and abetted, presumably, by an electronic diary or time manager that reminded him when my birthday was). As it happened he wasn't such a great accountant and this was the final straw. He had become a robot and I had become just a number. He was fired the old-fashioned way.

Telepresence and other warped ideas

Some people like to work on their own—at home, for instance—but physical interactions are more important than many people imagine. This could be a problem for organizations like Cisco, Hewlett-Packard, and Crayola that are promoting telecommuting, virtual meetings, and telepresence. Companies are even setting up offices in virtual worlds like Second Life to enable dispersed workers to communicate with each other.

The more time people spend outside the office, the less opportunity they have to physically share information or experiences. At IBM, for instance, 40 percent of its 386,000 employees have no permanent office, which has led to the moniker "I'm By Myself."

In theory this is a good idea. Corporations save time and money and there are some benefits to the environment too. However, something would be missing in a world where corporate HQs ceased to exist and all business was conducted virtually. One of the biggest downsides to mobile and virtual technology is the fact that it increases isolation. Sometimes being by yourself can be very productive, but I wonder what the longer-term social consequences of virtual living and virtual work might be. What will happen to us (and the quality of our thinking) if we choose to be surrounded with information and images rather than people and physical objects? What are the implications in terms of personal relationships, family life, or community?

It's a bit like paper versus pixels again. Both have their uses and the key is finding out what works best in each situation and for each individual. Nevertheless, I have a gut feeling that the hype surrounding virtual meetings, virtual laboratories, and e-learning will go much the same way as telecommuting did in the 1980s.

Part of the problem here is that large organizations tend to confuse efficiency with effectiveness. Efficiency is getting a specified job done correctly (on time and on budget with the minimum of mistakes). Effectiveness is ensuring that you are doing the right job. The latter requires deep thinking; the former does not. But firms are obsessed with superficial efficiency and short-term performance. Moreover, governments and corporations alike are addicted to public opinion, which can be fickle at the best of times.

Overall, organizations need to stop chasing opinion that has not been properly formed and start creating longer-term solutions. This, naturally, means that we must overcome our addiction to quick thinking. Companies the world over spend fortunes on schemes like Six Sigma, which aim to reduce errors and mistakes, but they spend next to nothing on improving the level of original thinking inside their organizations. Improving the quality of lateral or conceptual thinking is important but it's not urgent, so it never really happens.

In theory you'd think that all the time-saving technology we have at our disposal would have freed up more time to spend on activities like deep thinking, but the very opposite seems to be true. And it's getting worse. We are continually inventing new ways to make ourselves busy and appear occupied. I doubt whether Clay Shirky, author of *Cognitive Surplus: Creativity and Generosity in a Connected Age*, would agree with this. He claims that rather than passively watching sitcoms on television, we are now collectively creating civic value via the internet. He cites Wikipedia as an example, but beyond this and a few other examples I fail to see where the "creativity and generosity" of the book's subtitle lie.

The business of busyness

It's not just that we are intent on hurrying up our own thinking, these days we seem intent on hurrying up the thinking of everyone else too. It is, as James Gleick points out, as though the whole world has a Type-A personality and is suffering from hurry sickness.

Being busy has become a subtle social sign that we are important and successful. Hence we rush from one task or problem to the next without really stopping to think where we are going. Literally.

I know of one corporate strategist at a large bank who openly admits (to me at least) that his days are spent running from one meeting to the next. The purpose of each meeting is to discuss a set of documents that nobody has read properly beforehand because they are too busy. Hence they are scanned a few minutes before each meeting commences and everyone around the table puts on a poker face so that nobody guesses they haven't done their homework. Of course, everyone knows exactly what's going on. Most of the time this isn't a problem because the decisions are not critical. But every so often an expensive mistake gets made or a big opportunity is missed.

Another individual I've spoken to is a partner in a large law firm. He told me about a major mistake made by a young employee who missed a critical clause in a contract because he was trying to read it too quickly. Speed, it seemed, was the measure of all things for him. And I thought that lawyers were the only people left on earth who still read things properly. It's not only the legal eagles who are suffering: Lord Judge of Draycote, the Lord Chief Justice in the UK, has said that young people now make bad jurors because they lack the ability to concentrate for extended periods, especially when spoken information is presented as evidence in court.

No wonder, then, that a growing number of organizations are trying to reduce busyness by switching off email every Friday. Companies like Nestlé Rowntree in the UK and US Cellular in the US are encouraging email-free days or meeting people face to face rather than online. This is even happening in high-tech companies. US-based Intel, for instance, has told staff to phone or meet people physically whenever possible during zero-email Fridays.

Being busy prevents us from asking deep and difficult questions about ourselves. We do not like being alone with our thoughts any more than we like to be seen doing nothing. They are the twin terrors of our electronic age. A further modern malaise is anxiety, which is soothed by the illusion of control that we gain through constant technological connectedness. Writer Carolyn Johnson sums things up succinctly: "Distraction isn't merely available, it's unavoidable."

Organizations worry too much about what people are doing. If employees look busy it is assumed that they are being productive and vice versa. But these days many organizations are buying a person's ideas, not their time; or at least they are increasingly buying wisdom derived from experience. Hence less emphasis should be placed on superficial activity and more should be put on openness and receptivity. Less speed, in other words, or as Robert Grudin, author of *The Grace of Great Things*, puts it: "Stop searching for ideas and simply make room for them to visit."

Domestic thinking spaces

If we can free our mind of digital distractions, where does our thinking feel at home? Let's start with porches.

I doubt whether the average 10 year old in many parts of the world knows what a porch is, although a few might suggest that it's a German sports car. The decline of the porch can be laid at the feet of a number of well-known suspects: the development of new technology such as television, rising real estate values, or, most recently, the internet. This is a shame, because porches were great thinking spaces and porch talk had a depth and reflective quality that modern communications like email and instant messaging can't touch.

In the olden golden days people gravitated to porches to reflect on matters of both substance and triviality. They were fueled by the daily rhythms of small town life, but also by the serendipitous nature of fleeting conversation. As Philip Gulley, author of *Porch Talk*, observes:

> *"I do not wish to romanticize the porch. Not all talk reached the level of Plato or Jefferson but there was a luster to those talks, a certain glow... perhaps it was the parenthesis of silence, the bracketing of conversation with reflection."*

Moving inside the home, things have changed drastically here too. We have already seen how many organizations have torn down private offices and replaced them with varying types of open-plan seating or cubicles. Something similar has happened inside the home. It has been estimated that 2.9 million rooms have been lost in British homes since 2003 due to open-plan home conversions. This theoretically increases interaction between family members, although I suspect that, in practice, digital devices are getting in the way again.

A year ago there was an advertisement in Australia for Big Pond, an internet service provider. The ad spoke about "Big Pond homes" and showed a family of four people. Each was doing something different on a separate screen, in a separate area of the house. Hardly an advertisement for social cohesion.

However, it's not just the lack of private rooms. Private houses are also losing dining tables, landline telephones, and single television sets—areas and objects that forced individual household members to get together.

We have seen this before, of course. The piano, the gramophone, and the radio used to have much the same effect of bringing people together physically. These objects have now either disappeared or been replaced by personalized devices that allow the user to change what was once a communal experience into an individual one.

Architecture and urban planning have changed too. Terraced houses no longer tend to spill out into the street and, as a result, people nowadays live their lives with less direct contact with neighbors and passing strangers. These days, increasingly, we hide behind our walls, and casual contacts and conversations are severely limited.

But conversation is critical to ideas and innovation. Moreover, the very technologies that are bringing us closer together on one level are tearing us apart on another. Email, SMS, social networks, and the like are addictive attention eaters.

While we may be communicating with each other more, we may be listening and understanding each other less.

In *The Art of Conversation*, Catherine Blyth quotes some research claiming that the level of parental conversation in a household deeply affects the accomplishment of children outside the home many years later. Put another way, obsessive parental use of BlackBerries and PCs at home could retard a child's personal development.

A recent British government report also claimed that the increase in teenage pregnancies could partly be attributable to a decline in family conversation, because there was no longer the time, space, or inclination for parents to sit down as a family and discuss frank and meaningful topics with their children, such as the dangers of unprotected sex.

Conversational skills are not something you are born with, they are acquired. Historically they were acquired around a dinner table, but dining rooms are disappearing, being replaced in many cases with home offices and entertainment rooms. If you ask the typical screenager about anything beyond their immediate interests and experience, they generally give you a rather short and mumbled response. This might be language, but it isn't deep conversation—and it doesn't represent deep thinking.

Third places

Conversation isn't only suffering at home. "Third place" is a term coined by sociologist Ray Oldenburg to describe the intermediary places that we increasingly inhabit that are neither home nor work. Starbucks is a good example, as is the local pub or public library. These are also places where we can go to think, have conversations, and incubate ideas.

But have you tried to sit down and drink a cup of coffee in a

Starbucks in central London or New York recently? You may have found that all the tables were covered in laptops and spreadsheets, or all reserved for meetings. Indeed, the Coffee Office is a chain of cafés in Canada that offers mobile workers ("mobile warriors," in *Fast Company* magazine speak) somewhere to go when they have nowhere else to go. They look like normal coffee shops with free wireless internet and plenty of power plugs, but they also provide fee-paying members with workstations, conference rooms, meeting spaces, and sleep modules for power naps. They're basically a coffee machine with some offices attached, as opposed to an office with a few coffee machines.

Rather than being areas for relaxation, refreshment, and conversation, such places have become surrogate offices for people who don't have one or can't leave their work at work. Those described by another trendy term, the Bedouin worker, are in danger of destroying third places and, by default, turning them back into the second places from which people were trying to escape in the first place.

Unfortunately, trends such as delayering and downsizing are likely to continue and more people will be left with nowhere to go during the day. This won't last for ever, though. Spaces that outlaw laptops and business conversation will make a comeback. There are already bookshops and delicatessens that ban the use of cellphones. At the moment customers don't seem to take much notice, but eventually they will. I suspect that silent carriages on trains and quiet sections on airplanes are merely the beginning, too. Eventually we will see no-cellphone and BlackBerry-free sections of restaurants, as well as banning childcare solutions such as Sony PSPs and Nintendo DSs in certain public places.

At least, I hope that's what we will collectively decide to do.

Change your routine

Physical objects and environments are clearly important when we need to think. However, what is also important is that we break our routines from time to time. Most of us are immersed in fairly rigid daily schedules. We tend to go to the same places day in and day out, using the same routes and ending up meeting the same people.

The comedian Billy Connolly has a wonderful joke that the Queen of England thinks that the whole world smells of fresh paint. I suspect it's the same inside most organizations. Our worldview is colored by who we know and where we go and there isn't much variation or randomness.

There's nothing inherently wrong with this, especially if your objective is to explore a well-trodden path or to perfect something that already exists. We all know what practice makes and we cannot all be explorers and inventors.

However, the brain is constructed to respond to external stimuli and we are especially influenced by new physical stimuli. If the brain learns something new it grows; if it doesn't, it becomes lazy and starts to make short cuts. If new experiences are reduced or removed, our thinking starts to thicken or become less fluid and fresh. We start thinking in straight lines because this is the path of least resistance.

If you focus too much on a particular direction or fixed strategy, you can miss opportunities that crop up in a different or unexpected direction. Accidents are an essential ingredient of fresh thinking. You can sometimes learn far more from getting something wrong than getting something right, and clever ideas often flow from silly mistakes.

The list of major inventions created by accident or in unlikely locations is legendary. For instance, quantum electrodynamics popped into the mind of Nobel Prize-winning physicist Richard Feynman not at work but in a café in Ithaca, New York. The eureka moment was when someone was messing around with some plates and he saw a logo on the plate wobble. This gave rise to the idea of electron orbits. For another Nobel Prize-winning scientist, Kary Mullis, the eureka moment came driving down California Highway 128 in 1983. After 46.58 miles (scientists are nothing if not precise) Mullis suddenly had a thought about using oligonucleotides (short bits of DNA to you and me) to locate mutations in DNA. This eventually led to something called polymerase chain reactions. Of course, it would be less likely to occur today because Mullis would probably be distracted by a cellphone and automobiles are no longer the tranquil thinking spaces of yesteryear.

How about Post-It® notes? This idea came to 3M's Art Fry when he was singing in the choir loft at the North Presbyterian Church in St. Paul, Minnesota. The Harry Potter story occurred to JK Rowling during a car journey between Manchester and London. The concept of Netflix popped into the mind of Reed Hastings at the Canyon Video store in La Honda, California after he got hit with a $40 late return fee. The idea for nuclear chain reactions was conceived when Leo Szilard was crossing a set of traffic lights in Southampton Row in London.

We need to open our minds and feed them with juicy new bits of information, as well as nourishing them with random events and physical encounters. This isn't easy. It requires conscious and deliberate effort. Critically, it also requires slowness. Innovative thoughts don't generally come from a mind that's angry or hurried.

We should therefore think very carefully about what kind of thinking we are after and then think again about how to positively influence this kind of thinking.

But let's be clear about this. I am not calling for certain types of offices to be pulled down or for fast and convenient methods of communication to be abandoned. I am advocating simply that we should recognize what each object, tool, or environment is good at and seek some kind of balance.

10 ways in which objects and environments deepen our thinking

- ⟩ The overview effect: whether it is outer space, the top of a mountain, or looking out of a plane window, seeing an overview of the world inspires deep thinking.
- ⟩ Work may reduce deep thinking. External influences, such as sport or walking or literature, provoke more ideas than trying to have them at work.
- ⟩ Natural environments, like the ocean or gardens, inspire deep thinking because their rhythms slow us down and remind us that there are larger forces at work.
- ⟩ Changed office layouts—removing walls, creating proximity, open plan—are often not as effective as accidental or informal encounters and messy desks.
- ⟩ Telecommuting may be efficient and reduce paper and the space needed for offices, but it may also increase isolation and reduce the accidental encounters people and ideas need.
- ⟩ Bright walls and colored pens may reflect infantile culture more than stimulate ideas; clockless culture allows people to get the work done and find time for new ideas.
- ⟩ Brief encounters and accidents are great for deep thinking, but if you focus too hard in one direction you may not see something coming the other way.
- ⟩ Conversations at home, in the garden, on porches, or at the table stimulate deep thinking, but increasing use of technology is reducing the chances for conversation.
- ⟩ Third spaces (neither at home nor at work) provide a means of escaping the usual environment, although they can be overtaken by technology.
- ⟩ To think deeply or to have new ideas you need initially to be doing something superficially mundane or repetitive—you must think about nothing. Go somewhere new or do something different.

PART THREE

WHAT WE CAN DO ABOUT IT

Chapter 6
How to Clear a Blocked Brain

"Gentlemen, we haven't got the money, so we have got to think."

Ernest Rutherford

In the first two parts of this book we have examined how the digital era, and screen culture in particular, is rewiring our brains and how young minds are doing things differently to their forebears, sometimes with quite alarming results. We have also delved into how the human brain works, particularly with regard to deep thinking and the generation of new ideas, and looked at how specific external environments can influence how we think and what we think about.

This chapter outlines ten useful attitudes and behaviors for fostering creative, wide, and reflective thinking. This is not an exhaustive list, but it does offer practical advice about how to deliberately engineer deep thinking in the digital age. It is about how to balance analogue and digital behavior. These are tricks and tools that can be employed by anyone wanting to investigate their own thinking and unleash the creative potential of their mind.

My top ten ways of encouraging deep thinking are as follows:

- Create time and space
- Become intellectually promiscuous
- Keep an ideas diary
- Retain an open mind
- Use the bathroom
- Be patient
- Lose your inhibitions

⊼ Embrace failure
⊼ Share the problem
⊼ Don't go to work

Create time and space

> *"The time that you enjoy doing nothing is not wasted time."*
> Bertrand Russell

Time, as they say, is money, but we seem to have confused both the value of doing nothing and what we do. If people look physically active, we assume they are doing something worthwhile. But if they are sitting quietly and thinking, we assume they are wasting their time or ours. Daydreaming is seen as worthless procrastination and we overlook the value inherent in the slow flow. We need to develop unhurried minds.

To make the most of time we need to lose track of it sometimes. Buddhists talk about people with monkey minds: people who jump from one thought to another and are never wholly in the present or fully in the moment. We need to stop doing and just breathe. We need to cultivate mindfulness and this requires stillness. It is only when our minds are calm and empty and we stop thinking that the ideas come.

How can you do this? You have to train your mind through deliberate practice, but you need to recognize that this isn't something you can just do for twenty minutes before you need an idea.

In her book *The Innovation Killer*, Cynthia Barton Rabe highlights a survey by *Chief Marketing Officer* (CMO) magazine in the US finding that the biggest barrier to performance was insufficient time for strategic thinking and planning. This is scandalous. According to some 2008 research, the average office worker now receives 200 emails a day and we look at our inboxes

42 times a working day. This translates into 4 hours a day on technological interruptions. Moreover, the more senior you get, the less time you get just to think. No wonder books like *The 4-Hour Workweek*, *Getting Things Done: The Art of Stress-Free Productivity*, and others about managing attention sell like hot cakes. We need to ensure there is a commitment—one of what I call my 3Cs—to finding the time just to think.

The only way to solve this problem is defiance. This is obviously easier said than done in many organizations, because deep thinking isn't considered an important or high-priority activity. Not all organizations are like this, of course. 3M is famous for its 15 percent rule, which explicitly states that every employee is allowed to spend up to 15 percent of their time thinking about and developing new ideas. But it's not the 15 percent that's important. According to Geoffrey Nicholson, co-inventor of the Post-It® note, "Some people don't use that time; some people take more. It's the message that it's OK to dream."

Another benefit of allocating time is that people tend to work best when they are working on things that intrigue them and that they select for themselves. Google has a similar approach to 3M and allows employees to spend up to 20 percent of their time on personal projects that may or may not benefit the company. Gmail was one idea thought up when someone was doing something they weren't technically being paid to do.

So give yourself a break. It could be 30 minutes before work every Monday morning. It could be an hour-long walk every Friday afternoon, a proper long lunch with colleagues once a week, a regular massage, or a spot of meditation. It could even be a simple cup of coffee, but ideally not one from the office coffee machine.

I ripped an ad out of a newspaper a while ago that was selling "Essential espresso machines." The ad featured a testimonial from a satisfied manager called Dennis who stated, rather triumphantly: "Having the machine in our office saves a lot of time and money due to our staff being in the office instead of over at our local coffee shop." This irks me. The point of a cup of coffee is that it's a break from routine. It's a miniature moment of relaxation. You might even strike up an accidental conversation with someone at the local coffee shop that could take your mind somewhere interesting. Poor Dennis is missing something.

Perhaps instead of Googling yourself or sending Twitter updates on what type of coffee you are drinking at your desk, ditch the technological distractions and go to an art gallery, read a book, or draw a map instead. Or you could get really radical.

Imagine how powerful it would be for employers to give employees a break from their work every five years with no expectations or constraints and, ideally, the promise of their job back after the sabbatical. Gap years are generally the preserve of penniless students taking a break between high school and university or between university and work, but imagine if sabbaticals were readily available for everyone as a way of reenergizing ourselves or gaining a new perspective?

If it is difficult to physically remove yourself from work (or the home office), move mentally. Put some music on. Pour yourself a glass of wine. Yes, even in the office. I'm not suggesting that surgeons do this prior to performing an operation, but if you work in an office environment a glass of wine at lunchtime or one after work isn't going to kill you.

It doesn't really matter where you think—although as we saw in the previous chapter, some places are better than others depending on the kind of thinking you require—as long as you give yourself the time and mental space. Switch off any electronic devices if you can. Also remember that to have a really good idea you need to have lots of ideas, so you need to do this regularly. The volume of ideas you have comes back to the volume of deep thinking opportunities you give yourself.

It is worth remembering that most of your ideas won't be very good, but don't get hung up on this. It is important to get

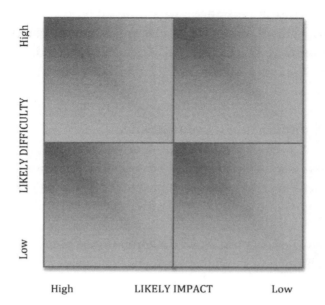

lots of ideas and eventually throw out the ones that are weak. If it helps, create three physical notepads, files, or boxes marked "no," "yes," and "maybe" and place a note of your thinking in the appropriate one. You could also consider using a matrix to sort your ideas according to likely impact (value) and likely degree of difficulty (practicality). Imagine a four-box matrix with an x-axis running left to right representing impact (high on the left, low on the right) and a y-axis running top to bottom representing practicality, difficulty, or perhaps cost of implementation (top being high, bottom being low).

You want ideas that cluster in the bottom left box or quartile (easy to do but high impact). You might also consider ideas that fall into the bottom right box (low impact but easy to do), because if you create enough of these tiny ideas, they may have a cumulative effect. Avoid ideas that fall into the top right box or quartile (things that are difficult to do and have low impact). The only proviso is that you must think very carefully about whether something really is high or low impact or high or low difficulty. In some instances, what initially looks easy will turn out to be difficult and so on. Sometimes it can make sense to do something that is very difficult because there won't be very many people following you.

Become intellectually promiscuous

"Think left and think right and think low and think high. Oh, the thinks you can think up if only you try."

Dr. Seuss

If you show a newborn child the same things their eyes start to wander. But if you show them something new their gaze returns. It's similar with adults. Most of us are trapped in routines. We travel the same route to the office, we sit with the same people, and we hang out with the same old ideas. Don't.

As we grow older it becomes increasingly difficult to move beyond what we have already experienced. So if you are after new thinking you must consciously disrupt routines and introduce fresh information and experiences. You must develop new neural paths, what Gregory Berns, author of *Iconoclast*, describes as "experience-dependent categorization." Put another way, you need mental stimulation.

It's not only individuals who get stuck in routines. Institutions tend to specialize, so they tend to know more and more about less and less. Moreover, individuals with certain functions, interests, or skills within organizations are usually physically grouped or clustered in fixed departments or locations. There is nothing wrong with this if you want to continually improve what you are doing. In this case specialization and best practice benchmarking are exactly what you should be doing. But if you want revolutionary, not evolutionary, thinking, you don't need more expertise but more intellectual promiscuity. Expose yourself to new experiences and find new pieces of disconnected information. Go on fact-finding missions. Walk around deliberately looking for new things and talking with new people.

How do you do this? Take an unusual route to work one morning. Buy a magazine you've never read before (one that isn't connected to your current work or interests), go on holiday somewhere new, start talking to people you don't know when you're on an airplane. Listen to music you've never heard before. It may seem like a waste of time, but I promise you that buried deep inside these disconnected experiences will be the seed of a new idea.

Harvard psychologist Howard Gardner says that high achievers can often be identified early by their love of topics, tasks, and issues that are not core or essential. The link here is with curiosity or, more specifically, a love of serendipitous experiences. There is plenty of evidence to show that within organizations the best ideas come from people who have the most contacts outside their immediate division or department.

On a practical level, if you are looking for ideas, ensure that any teams you have responsibility for are multidisciplinary. Involve customer service, customer complaints, sales, finance, and production. And whatever you do, try not to put together a group where everyone is the same age, the same sex, and went to the same school. Why not? Because they've had similar experiences and they'll all have the same ideas.

Ideas don't come out of nowhere. All of them are weakly connected to existing ideas. Therefore it's about cross-pollination, creating new combinations, new connections, and new neural pathways. So if you're interested in giving birth to new ideas, you first need to connect with other ideas through indiscriminate relationships with people, places, and things. Your ideas need to start having casual sex. It is from this frenzied coupling that new ideas will be conceived.

You might also try looking at familiar objects and environments with fresh eyes. Delving into the deep history of things can help, or interrogating ideas until they go back to their roots and confess where they've come from. Always start from scratch if at all possible and demand rigorous clarity in terms of the problem at hand. If the problem isn't clearly articulated, don't start thinking about potential solutions until it is.

Keep an ideas diary

"It is not a bad idea to get into the habit of writing down one's thoughts. It saves one from having to bother anyone else with them."

Isabel Colgate

Don't always rely on your subconscious to remember what you come across and spit out intriguing new combinations of ideas. Write any interesting thinking down (things you think, things people say, things you see) and be prepared to do this absolutely anywhere.

I'm not suggesting that you install a waterproof whiteboard in the shower (although an American hotel chain once did exactly that, based on research suggesting that this was where their guests had ideas), but keep a notepad by your bed, open a file on your computer... anything. Remember to revisit these ideas on a regular basis and avoid the temptation to throw your thoughts (or your notebooks) away too quickly. Sometimes a stupid thought becomes sensible over time. And remember what Einstein said: "If at first the idea is not absurd, then there is no hope for it."

Otto Loewi was a German biologist who won a Nobel Prize in 1936 for an idea he had in his sleep. Being alert to the power of his dreams, he awoke one night to write an idea down, but unfortunately the following morning all he found was an incoherent scribble on a pad beside his bed. If his idea was "buy apples today" it wouldn't have mattered much, but he instinctively knew that his thought was important.

Fortunately, the idea paid another visit the following night and this time Otto was sufficiently composed not to let it escape. He got dressed, wrote the idea down, and went straight to work in his laboratory. This ultimately led to the discovery of acetylcholine and Otto became the father of neuroscience. Not bad for an idea he had in his sleep.

Again, how you capture ideas is down to you. Some people prefer to use pen and paper, some use a voice recorder, and others a camera, if it's a visual idea they want to preserve. I haven't seen any research to suggest that the method of capture in any way influences the idea, but small bits of paper could obviously be unreliable.

As to what other people will think of you furiously scribbling things down in the middle of the night, during meetings, or halfway down the vegetable aisle in the supermarket, get over it. That's their problem, not yours. Whipping out a notebook in the middle of a funeral might raise a few eyebrows, but if you feel an idea is important enough, scribble it down.

Retain an open mind

"There exist limitless opportunities in every industry. Where there is an open mind there will always be a frontier."
Charles Kettering

Dawna Markova, in her book *The Open Mind*, discusses how education teaches us that there is a singular answer to everything. Rather than having an open mind, we are cajoled into deciding the right answer in any given situation, and this inevitably kills off all other possibilities.

I've come across this tendency myself, working with large organizations that seek truth in the sense of a singular, perfect solution. Things start off well enough. Research is conducted, insights are uncovered, and this eventually leads to countless alternative ideas being discussed and some being developed. Next there is a rush to find the best idea, be that a new strategy or a new product concept. Ideas fall by the wayside, either because they are not supported internally, or because a statistically representative sample of people (quite literally picked off the street) "can't see it."

I bet you a zillion dollars there will be several ideas of more or less equal merit and if you keep an open mind, you can just pick one of them and implement it as rapidly as possible. Small adjustments can be made after launch. If the idea is strong enough, you can always water it down later, as long as you execute it well. But if the idea is weak, you haven't got a hope.

Organizations tend to have a predetermined view of what will work and then reject any idea that does not fit with this viewpoint or lens. Cordelia Fine speaks of the brain behaving like a corrupt detective searching for evidence to support the view that an individual is guilty. As she puts it, our subconscious brains have usually already made up their minds "by hiding or destroying files that harbour unwanted information."

We maintain the illusion of control when all the evidence suggests that we control next to nothing. We think (briefly), we plan (poorly), we strategize (for the next 18–36 months out), but most of this is superficial and easily disturbed. Hence not making up our minds until the last moment—or having a plan that is relatively fluid—is actually not a bad strategy. But once we have made up our minds, it is very difficult to unmake them. The windows of perception are slammed shut. To quote Fine again: "So long as our minds are yet to be made up, we actually view ourselves and life unusually realistically as we quietly contemplate our future."

William Wordsworth had much the same idea. He said: "Not choice but habit rules the unreflecting herd." But don't forget that the brain is always open to new experiences. Thus, it's possible to create new superhighways of information (new synaptic paths and even new brain cells) that can bypass your old freeway networks.

Consciously develop new habits. Think of yourself as having three levels of comfort: a comfort zone, a stretch zone, and a stress zone. Most of us spend most of our days doing what we always do and consequently our learning rate is practically zero. But if we try to do new things on a regular basis (and formalize it so that it always occurs) we stretch ourselves. And, as we know, when a mind is stretched it rarely returns to its original dimensions. Again, this links to one of my 3Cs, in this case the second C—curiosity.

Another way of keeping an open mind is to be a bit naïve from time to time. A certain degree of innocence can be useful, even commercially. Don't be afraid to ask seemingly naïve questions, even if you already know the answer.

I occasionally use a trick called the "seven levels of why" (which I borrowed from a friend, Andrew Crosthwaite) to get to the bottom of why things are as they are. As Andrew puts it (and I've stolen some of his best words with permission), when we are children we are incessantly asking our parents or teachers "Why?" It's one of the ways we learn. But when we get older we ask the question less frequently. This is because our apparent need to ask why diminishes, but it's also because we feel faintly embarrassed to ask the question. We assume that we should know. So we don't ask out of fear of embarrassment, or we feel that the organization already knows so it is unnecessary to ask. We get on with other things, safe in the knowledge that everything that needs to be known has already been found out.

Lazy assumptions should always be challenged when you see them. This doesn't need to be destructive, it is simply using investigative techniques to uncover how things really work. Think of it as an argument map or a decision tree process. Look at what you are doing or thinking and throw down the "Why?" gauntlet. Do this every so often because if you don't do it, a competitor or a customer inevitably will. Where will this questioning ultimately take you? I can't say, but I suspect that you'll be pleasantly surprised where you end up, not least because what you don't know always outweighs what you do.

Innocence can be tapped in other ways too. You've organized yet another focus group to figure out how people buy new cars. Or perhaps you've put together a brainstorm to develop ideas for a new car. You've got a room full of the usual suspects and, as a result, you get all the usual insights.

This approach is next to useless if you want to uncover some really deep insights or develop something entirely new. And that is where another thinking exercise comes in, "innocent experts." I've been using this tool for many years and it seems to capture people's imagination. Instead of running a focus group with "ordinary" people, jump into a car with a series of extraordinary people and visit a few auto dealers unannounced.

This is something I did with a Japanese car company many years ago. The unusual suspects included an architect, a product designer, an anthropologist, a psychologist, an MBA student, an insurance salesman, a supermarket space planner, and an assortment of iconoclasts and revolutionaries. Half of them were men, half were women, and their ages ranged from around 20 to more than 60. On another occasion I put together an eclectic group on behalf of a chemical company that was interested in the trends driving paint colors and finishes, only this time the meeting of minds was at dinner. Why dinner? Why not? In my opinion, the food and the wine (and the physical space) directly influenced the quality of the debate and the thinking was significantly better than if we had all met in a conference room.

What are the practical benefits of such mental melting pots? In a group that's diverse in skills and experience you get a high level of cross-fertilization, giving you fresh perspectives. People also enjoy meeting those outside their own discipline, which again sparks the conversation. Most importantly of all, outsiders can be experts at seeing things that aren't there, because they don't yet exist. Because innocent experts are outsiders, they have no knowledge of what's politically correct. Thus, they tend to ask questions that are fundamental. They have no knowledge of the rules and are therefore quite fearless.

For example, one innocent expert (a historian) asked a water utility company what business it was in. The client (the head of R&D) was at first insulted by the question. But eventually he realized that it was a catalyst for thinking about where the company could go in the future.

Use the bathroom

"Genius is editing."

Charlie Chaplin

According to researcher Nikolai Shevchuk, modern life is lacking in physiological stress factors and, as a result, our brains are not as sparked up as one would ideally like them to be. His research claims that short, sharp shocks in the form of cold showers stimulate a part of the brain known as the blue spot (*locus ceruleus*). This is the main source of a chemical called noradrenaline that scientists believe is linked to depression. Give this region a shock with a three-minute blast of very cold water and our mood can change considerably, snapping us out of negative thoughts and helping to alleviate chronic stress.

I suggest that you check with your doctor before trying this out, especially if you have a heart condition, but maybe the old

English public schools (and the ancient Chinese) were right all along. Cold water really does work wonders for altering your mood, which in turn affects how you think.

The writer Tom Hodgkinson has a few things to say about the modern era that I wholeheartedly agree with. For example, he thinks it important "to read decent stuff. Put quality materials into your mind, quality ingredients." So if you are lying in a warm bath, read something good. He adds: "Do not go to the gym. Gyms are all mixed up with vanity and money, with the absurd quest for perfection. They are the consumer ethic transferred to the body. They are anti-thought and their giant screens blot out our minds and divert us from ourselves."

There is a bit of a jump from bathrooms to gyms, but I do think there is a connection. Gyms, for me at least, are all about relaxation; so are bathrooms. Both are also connected to well-being. For instance, research has clearly shown that exercise improves some of the brain's higher functions (such as planning) because it increases blood flow, which in turn improves the delivery of oxygen and nutrients to your neurons. Second, monotonous exercise, such as running on a treadmill, clears the mind to think of other things, as does lying in a bath or standing in a shower.

However, the digital era is invading even here. You can now buy waterproof televisions for shower cubicles and telephones are becoming quite common in bathrooms. Screens of various kinds are omnipresent in the gym.

So to unblock your brain, don't stare at yet another screen, take a long bath, have a long shower, sit on the lavatory. Do anything, in fact, that gets you away from the digital era for a few minutes and involves a certain amount of relaxation.

Be patient

"The two most powerful warriors are patience and time."
Leo Tolstoy

One of the biggest problems with big problems is that people give up far too soon. We think about things, we think a bit more, and then we quit—because nothing appears to be happening or because the process is frustrating or confusing. Yet a truly original idea can take years to arrive and successfully implementing it often requires decades of energy, passion, and enthusiasm.

Generating ideas or solutions tends to proceed through three clear stages: education, incubation, and illumination.

The education stage is deeply demanding. You need to think, a lot, and expose yourself to varied inputs. You need to be become conscious of the issues and sensitive to what is occurring around you. You need to listen deeply and watch carefully too. In short, you need to become receptive and focus your attention on the problem at hand.

Relaxation—a sense of mental calm and physical quiet—is essential during this stage. Judging from a host of recent experiments, when you're highly stressed you desire and require fast information. Indeed, fast facts help you to relax. However, once you are relaxed, fast-paced information can undo the state of calm and make you stressed again.

The incubation stage is unnerving, largely because you don't know you're in it and you cannot directly control it. However, to think that nothing is happening would be a colossal mistake. Your brain is working overtime during this stage and there's plenty going on, it's just that all of the work occurs in your unconscious. Your mind is working away and the vital point is to wait. Tenacity is crucial. Endurance thinking, someone once called it.

Eventually (the illumination stage), something will pop into your head, usually unannounced. At this point, the flow of ideas

THE BENEFITS OF BOREDOM

Boredom is beautiful. Rumination is the prelude to creation. Not only is doing nothing one of life's few remaining luxuries, it is also a state of mind that allows us to let go of the external world and explore what's deep inside our head. But you can't do this if ten people keep sending you messages about what they are eating for lunch or commenting on the cut of your new suit. Reflection creates clarity. It is a "prelude to engagement of the imagination," according to Dr. Edward Hallowell, author of *Crazy Busy*. It is a useful human emotion and one that has historically driven deep insight.

Boredom hurts at first, but once you get through the mental anguish you can see things in their proper context or sometimes in a new light. Digital technology, and mobile technology in particular, appears to negate this. If you are trying to solve a problem it is now far too easy to become digitally distracted and move on. But if you persist, you might just find what you've been looking for. So don't just do something after you've read this chapter, sit and think for a while.

Faced with nothing, you invent new ways of doing something. This is how most artists think when faced with a blank canvas. Historically, children have operated like this too. They moan and groan that they are bored, but eventually they find something to do—by themselves. Boredom is a catalyst for creative thought.

Only these days it mostly isn't. We don't allow our children the time or the space to drift and dream. According to the UK Office of National Statistics, 45 percent of children under 16 spend just 2 percent of their time alone. Moreover, the amount of free time available to schoolchildren (after going to school, doing homework, sleeping, and eating) has declined from 45 to 25 percent. Children are scheduled, organized, and outsourced to the point where they never have what New York University Professor Jerome Wakefield calls a chance to "know themselves." It's the same with adults. Our minds are rarely scrubbed and dust builds up to the point where we can't see things properly.

Not only is it difficult to become bored, we can't even keep still long enough to do one thing properly. Multitasking is killing deep thinking. Leo Chalupa, an ophthalmologist and neurobiologist at the University of California (Davis), claims that the demands of multitasking and the barrage of aural and visual information (and disinformation) are producing long-lasting and potentially permanent damage to our brains. A related idea is constant partial attention (CPA). Linda Stone, who has worked at both Apple and Microsoft Research Labs, knows about how high-tech devices influence human behavior. She coined the term CPA to describe how individuals continually scan the digital environment for opportunities and threats. Keeping up with the latest information becomes addictive and people get bored in its absence.

In a sense this isn't anything new. We were all doing this 40,000 years ago on the savannah, tucking into freshly killed meat while keeping keeping a look out for predators. But digitization plus connectivity has increased the amount of information it's now possible to consume to the extent that our attention is now fragmented all of the time. This isn't always a bad thing, as Stone points out. It's merely a strategy to deal with certain kinds of activity or information.

However, our attention is finite and we can't be in hyper-alert, "fight-or-flight" mode 24/7. Constant alertness is stressful to body and mind and it is important to switch off, or at least reduce, some of the incoming information from time to time. As Carl Honoré, author of *In Praise of Slow*, says: "Instead of thinking deeply, or letting an idea simmer in the back of the mind, our instinct is now to reach for the nearest sound bite." We relax by cramming even more information into our heads.

Chalupa's radical idea is that every year people should be encouraged to spend a whole day doing absolutely nothing. No human contact whatsoever. No conversation, no telephone calls, no email, no instant messaging, no books, no newspapers, no magazines, no television, no radio, and no music. No contact with people or the products of other human minds, be it written, spoken, or recorded.

Have you ever done nothing for 24 hours? Try it. It will do your head in for a while. Total solitude, silence, or lack of mental distraction destroys your sense of self. Time becomes meaningless and recent memories start to disappear. There is a feeling of being removed from everything while being deeply connected to everything in the universe. It is fantastic and frightening all at the same time. But don't worry, you soon feel normal again. Return to sensory overload and deep questions about a unifying principle for the universe soon disappear, to be replaced by important questions about what you're going to eat for dinner tonight or how you're going to find that missing Word file.

Consider what Bill Gates used to do. Twice a year for 15 years the world's then richest man would take himself off to a secret waterfront hideaway for a seven-day stretch of seclusion. The ritual and the agenda of Bill's think weeks were always the same: to ponder the future and to come up with a few ideas to shake up Microsoft. In his case this involved reading matter but no people. Given that Gates has been instrumental in the design of modern office life, it's interesting that he felt the need to get away physically; one would expect him to inhabit a virtual world instead.

I once received a brief from the strategy director of a FTSE 100 company who wanted to take his team away to do some thinking. When I suggested that we should do just that—go away for a few days, read some books, think, and then discuss what we'd read—he thought I'd lost my mind. Why? Because there was "no process." There were no milestones, stage gates, or concrete deliverables against which he could measure his investment.

The point of the exercise is this. Solitude (like boredom) stimulates the mind in ways that you cannot imagine unless you've experienced it. Solitude reveals the real you, which is perhaps why so many people are so afraid of it. Empty spaces terrify people, especially those with nothing between their ears.

But being alone and having nothing to think about allows your mind to refresh itself. Why not discover the benefits of boredom for yourself?

can often turn into a flood, when all of the individual elements start to join together in a frenetic frenzy.

The danger here is that, because stage three feels effortless, we don't work enough on stage one or allow enough time for stage two. We assume that ideas will just float into our minds, so we become lazy and don't work hard enough at the first stage, or we give up too soon because there's no guarantee that anything useful is going to happen in stage two. The ground needs to be properly prepared before anything big can grow.

When asked how he discovered gravity, Newton simply responded that he'd thought about it a lot. Brilliant! Self-doubt, fear, confusion, frustration, and inadequacy are all part of this process, but you should keep calm and carry on regardless. Embrace both uncertainty and boredom with a wry smile whenever they appear, because they are the forerunners of change. Equally, don't berate yourself too hard when nothing happens. Keep at it and eventually something will.

The harder you work, the better the ideas become too. For example, when Einstein was working on his theory of relativity he was often depressed and at times in despair. Indeed, his questioning and thinking went on for a total of seven years, although his final theory evolved over a period of only five weeks.

Waiting and being patient also link with meditation, in the sense that being relaxed appears to have a positive impact on mood and thus to increase focus and attention. The Dalai Lama once stated: "Solutions never come from an angry mind."

Patience is crucial, so don't be in a rush or give up too early. Big ideas and small ideas are part of the same linear continuum, a point developed by Ronald Beghetto at the University of Oregon and James Kaufman at the California State University at San Barnardino. Giving up because your ideas aren't good enough (quality) or because you haven't had any or enough of them (quantity) misses the point. You can never know in advance whether an idea that flashes into your mind will be a good one or a bad one. The best strategy is simply to keep having ideas until you have one that is big enough for you.

Lose your inhibitions

"Neck ties strangle clear thinking."

Lin Yutang

Are you sitting comfortably? Then I'll begin. It is possible that you can germinate or propagate a great idea when you're busy or stressed, but it isn't very likely. A relaxed environment is usually more conductive of deep thinking. The best environment is somewhere where you are conscious but not self-conscious.

Where, for example, are you reading this book? Are you sitting or lying down? I'd wager that you're doing one or the other, rather than standing up. Your body needs to be relaxed for your mind to properly connect with the information and ideas within the book and to allow you to reflect and contemplate.

In case you are reading this standing up (there's always one) and you're wondering why I haven't noted that Winston Churchill, Ernest Hemingway, Virginia Woolf, and Thomas Jefferson all worked at stand-up desks, I'd answer thus. Yes, the idea of a stand-up desk is a good one. They make meetings shorter (hurray), they burn more calories (seriously), and they solve many of the neck and back problems encountered by sedentary desk jockeys. But standing up isn't a good way to think deeply. To soak up a good book (or ponder a big problem) you need to be relaxed and you need to lose your inhibitions. To do this you need confidence, the third of my 3Cs.

A study by psychologists at Harvard University and the University of Toronto has claimed that highly original thinkers tend to have "low latent inhibition." Latent inhibition is a useful trait that allows us to filter out unwanted information. For instance, if you took notice of every single sound you heard today you'd go slightly mad. Some things just aren't important (the sound of a printer warming up), although the sound of breaking glass or a scream might be. Highly original thinkers tend not to file or classify information as tightly as others. Thus there is more room inside their heads to make connections or to notice things that others have missed. Clearly, getting the balance right is important, but letting in more information, rather than less, would appear to be a good idea.

Losing your inhibitions is important in another way too. You need to be relaxed about not being judged (a link to childlike thinking) and take criticism in the right spirit. You also need to remove your guard to let in seemingly silly, unconnected, or crazy questions into your head. As a result, you need to be in a physical environment that is nonjudgmental.

Another way of losing your inhibitions is to listen to music, since this has a direct effect on your brain. Pick the right music for you and the good vibrations are likely to activate your brain's reward centers and simultaneously depress the amount of activity in your amygdala, to lessen fearful feelings and other negative emotions.

Embrace failure

> "I have not failed. I have just found 10,000 ways that won't work."
>
> Thomas Edison

You don't read about failure very often. I'm not just talking about great ideas that never see the light of day, I'm talking about people too. Most companies—in fact, most people—fail far more often than they succeed. This is a proverbial elephant in the boardroom. But by being so scared of failure, we are missing a valuable opportunity.

The point about failure is not why it happens but what we do when it does. Most people run. Or they find a way to be "economical with the actualité," as a former British government minister once described it. "We launched a counteroffensive too late." "Our consumers weren't quite ready for it." No. You failed. Own up to it. It's the beginning, not the end.

The problem is simple: Most people believe that success breeds success. But to me, failure breeds success. There are plenty of people who fail before they succeed and some are serial failures. Indeed, an American investment banker friend once told me that there is a venture capital firm in California that will only invest in you if you've gone bankrupt at least once.

James Dyson wasn't funded by this firm, but I'm sure they would applaud his spirit. The billionaire inventor of the bagless vacuum cleaner built 5,127 prototypes before he found a design that worked. But he looked at his failures and learned. He never made the same mistake twice. Each adaptation led him closer to his goal and out of the chaos and uncertainty came success. As someone once said (I think it was in the children's movie *Chitty Chitty Bang Bang*), there's magic in the wake of a fiasco. It gives you the opportunity to second-guess. Failure gives you the opportunity to think. Again. As Isaac Asimov commented: "A subtle thought that is in error may yet give rise to fruitful inquiry that can establish truths of great value."

None of this is to be confused with the mantra of most motivational speakers who urge you not to give up. Success is 1 percent inspiration and 99 percent perspiration, they say, and if you just keep on trying for long enough, it will eventually happen. And if it doesn't, you're just not trying hard enough. Doing the same thing over and over again in the hope that something will change is one definition of madness. What you need to do is learn from each failure and try again *differently*.

Critically, it is what you do when you fail that counts. Remember Apple's message pad, the Newton? Probably not. It was a commercial flop, but the failure was glorious. The tolerance of failure embedded in Apple's DNA may be one of the reasons for its success with the iPod, iTunes, and the iPad. And who could have guessed that the one-time AIDS wonder drug AZT had been a failed treatment for cancer, or that Viagra was a failed heart medication that Pfizer stopped researching in 1992?

To quote Italian designer Alberto Alessi: "Anything very new often falls into the realm of the not possible, but you should still sail as close to the edge as you can, because it is only through failure that you will know where the edge really is."

The edge is where real genius resides. It can be a difficult location, especially when you stop to consider that the bigger an idea is, the more opposition it will encounter. A truly visionary idea will, by its very nature, threaten whole structures of understanding.

English sculptor Henry Moore sums this up: "The secret of life is to have a task, something you bring everything to, every minute of the day for your whole life. And the most important thing is: It must be something you cannot possibly do." How's that for a definition of embracing failure?

Share the problem

> *"Given enough eyeballs, all bugs are shallow."*
> Eric Raymond (about open source software)

Organizations used to create departments to create ideas. Some still do. But what tends to happen is that idea creation or development is turned into a state secret. The only people who know what is going on, and thus the only individuals who can contribute, are the Chosen Ones.

This is crazy, because it limits both the quantity and the quality of ideas. As a result, smart firms have started to play around with new ways of developing new ideas. One such concept is distributed or open source innovation, in which customers (or everyone else, for that matter) are the co-producers of the products and services they consume. Innovation works best when it's bottom up rather than centrally planned. As we've seen, group intelligence has its limits, and groups are better used to solve well-defined problems rather than to find problems or invent new ideas, so it's important to find the right balance between collective and individual thinking. But, generally speaking, the more minds you put on a problem, the faster the problem shrinks.

This idea began with software. The open source movement worked there because the original motive was highly altruistic: the end product was given away for free, and people thought they were on the side of David fighting a mighty Goliath (or Bill). Networked innovation is made possible by the internet. Ideas (or problems for that matter) are freely available to anyone who's interested.

Recently the concept has been transferred to all manner of projects, ranging from open source encyclopedias to airplane design, cola recipes, and film scripts. One brand of beer was developed with the help of some self-appointed aficionados (found on the internet) who created everything from the name of the beer to its packaging and advertising. Even NASA has embraced the idea by using volunteer scientists (or click workers, as they have become known) to identify and catalogue craters on the surface of Mars.

In some ways open source innovation can be thought of as a giant suggestion box scheme, albeit one with a transparent box. A problem is posted on a website and anyone from industry experts to members of the public can contribute to the solution. Everything is transparent as all ideas are shared and discussed in public. In some instances people will do this for nothing, while others have to be paid in some way.

Some highly vocal detractors argue that open source innovation is little more than a giant focus group, but there are differences. The first is sheer scale. Focus groups rarely involve more than 100 people. Open-source innovation can involve thousands and still turn things around faster than more traditional approaches. Second, focus groups usually ask people to react to ideas. Open source asks people for solutions and allows ideas to build cumulatively. Third, focus groups rely on a representative sample of people who are "ordinary" and by definition uninterested. Open source relies on self-motivated people who are articulate, passionate, and enthusiastic.

John Kao (*Jamming*) and Ross Dawson (*Living Networks*) argue that the traditional innovation model can be compared to classical music, whereas the open source model is more like jazz. Classical innovation is usually driven by a single leader with a team following detailed notes. The end result is faithful in letter and spirit to the original objective. Open source, in contrast, involves individuals improvising against a background score. Sometimes there is no leader or set framework and the results can be quite unexpected.

A good example of a company moving toward an open source approach is Procter & Gamble. It has an objective to generate 50 percent of new product ideas from outside the company. Its Collaborative Planning, Forecasting and Replenishment (CPFR) is a transparent process that allows its customers and suppliers to improve its supply chain. Another example is P&G's use of the virtual technology market yet2.com. P&G lists every one of its thousands of patents on yet2.com in the hope that this will facilitate connections and ideas from the outside.

Most open source innovation projects are total failures, but this is precisely why the concept is so important. Within traditional innovation models the cost of failure is very high. As a result, stage gates, red lights, green lights, and funnels are introduced to control the number of new ideas that are developed and introduced to the market. And that's a big problem. Nobody ever knows for sure what will work and what won't, until an individual (or an organization) makes a leap of faith and an idea is unleashed. Nobody can tell what's silly and what isn't without the benefit of hindsight. But with open innovation, researching and worrying about whether something will work or not is unnecessary because of the low cost of trying. Every employee, customer, and stakeholder is facilitated and empowered to become part of the innovation team and then to perform small experiments.

Of course, the idea of setting up some kind of anything-goes system would be anarchy. What is needed is a balance, a combination of tight and loose, where 85–90 percent of resources are spent on innovation that is tightly planned and controlled. The remaining 10–15 percent of time and money should be invested in unplanned ideas that are developed by simply releasing them into the wild and seeing what happens.

How can you use the open source model to kickstart innovation within your organization? One suggestion is to clearly articulate a problem and then ask for well-expressed solutions from anyone who's interested, inside or outside the organization. Another is simply to seek out passionate enthusiasts, such as heavy users of your product or service, and then collaborate with them.

One final thing connected to openness. Intuition is often dismissed because it isn't scientific. But quite often intuition is actually our subconscious mind poking through. Using tumbling and contradictory intuition can sometimes take us to places that logic can't. Be open to it.

Don't go to work

"The quality of the imagination is to flow and not to freeze."
Ralph Waldo Emerson

If you find that your mind is getting stuck, one of the best things you can do is distract yourself. Simply take the day off or go on holiday. Nevertheless, these simple acts seem to be getting more and more difficult.

The term "leisure guilt" was coined by American psychologist Raymond Folen. Overworked individuals become enveloped in anxiety, so when the opportunity to take a week off emerges, they find reasons why they can't go. Essentially, we now value work so highly over most other activities that we're made to feel guilty about not working.

Leisure guilt—more of an idea than a condition—means that people don't take vacations or that they virtually sneak into the office while on holiday (via BlackBerries, cellphones, and email). Employees claim they're being conscientious, but their actions represent a deep level of insecurity: fear of being fired or getting "out of the loop." This perpetuates a cycle of burnout and fatigue.

Some companies are on to this problem. For instance, Boston Consulting Group has adopted a limited hours policy for all staff in the US that explicitly states that hours worked have nothing to do with future promotion. Moreover, the firm red flags employees who work more than 60 hours a week over any five-week period. I have yet to hear about an organization that makes holidays compulsory, but I'm sure there's at least one out there somewhere.

So become deliciously disconnected for at least an hour or two once in a while. Turn your cellphone off or reset your BlackBerry so that it can send but not receive email. Separate work from pleasure with different communication tools or social networks. Above all, become an information cynic. Do not treat all incoming information as urgent or important.

If your mind has gone down a dead-end street, back up, stop the car, and get out. Go for a walk. Above all, don't go to work. Hardly anyone I spoke to over the course of writing this book mentioned the office as somewhere they had their best ideas.

Chapter 7

Future Minds

"I may not have gone where I intended to go, but I think I have ended up where I intended to be."

Douglas Adams

I said at the beginning of this book that deep thinking is important because it changes things. It makes the world a better place. But there is another reason why it matters. Deep thinking is personally fulfilling. It is our deep thinking that makes us uniquely human. It is deep thinking that gives rise to our awareness of self and for this reason alone we should cherish and nourish our minds at every conceivable opportunity, even when it appears inconvenient or expensive.

Control-Alt-Delete

Critically, this means that we need to do less and think more. It means slowing some things down a little. It means occasionally valuing intuitive and creative thinking over thinking that is rational, analytical, and logical. It especially means that we need to get away from the idea that all communication and decision making must be done instantly. It means that we must fight against the convenience of the digital era. In the words of Jack White from the White Stripes: "That's the disease you have to fight in any creative field—ease of use."

It also means that we need to recognize that deep thinking can't be done in a hurry and that certain kinds of learning cannot be done in fast-paced environments or with the aid of digital tools. Virtual learning? A virtual teacher is about as much use as a virtual parent or a virtual pet. Content is always important, but so too are the method and speed of delivery. If we fully embrace a process that makes things happen too quickly, we are prevented from properly analyzing or discussing things. There is just no time for clarifying thought.

Ultimately, the question of whether or not digital technology is influencing how we think is misleading because, as we've seen, there is more than one kind of thinking, much as there is more than one kind of intelligence. Thus, the debate is not really whether there is any influence, but rather how certain types of digital technology affect the kinds of thinking we are most interested in.

Cellphones are a particularly pernicious problem. There are 4 billion of the little blighters and the number keeps growing. I'm not suggesting that we get rid of them altogether, because they are useful. But they can invade private spaces. They can adversely affect the quality of our thinking and our manners. Devices like iPhones and BlackBerries invite (demand!) constant use. They are like packets of cigarettes that ask us never to leave them alone or bottles of pills that seek to change our minds and punish us when we try to withdraw from them. We think cellphones are connecting us, but they are turning us into a society of rude, impatient, narrow-minded, stressed-out, aggressive, and isolated individuals. They invite reaction rather than reflection. Google, Facebook, and Twitter are doing much the same. They are all screaming at us, making us lazy and unempathetic. And, most importantly, they are interrupting the physical intimacy that both people and ideas require.

I am not declaring war against digital technology. It isn't intrinsically evil. I am simply saying that there are times—and places—where people, not machines, should come first. The digital era is turning spaces and activities that were once largely communications free, calm, and relatively silent into noisy extensions of offices and shops, none of which is particularly friendly to deep thinking. And my main concern is that computers are supposed to be tools to help us to think, not prevent us from thinking.

What computers are especially good at, what they were invented for, is processing large amounts of information, thus uncovering the insights that allow human beings to think deeply, solve problems, and have new ideas. But we have started to use tools such as these to replace or outsource our own thinking. Similarly, email and cellphones are useful ways of enriching physical conversation and contact, but we are starting to use them to replace it.

It is this replacement that concerns me the most. Given the fact that we seem to be capable of inventing more or less anything these days, perhaps a question we should be asking ourselves more frequently in the future is not whether we *can* invent something but whether we *should*. What are the potential consequences of some of these inventions? Indeed, there is a possibility that we are accidentally destroying many of the things that, deep down, we value the most.

Manmade machines cannot provide the one and only thing that we are all looking for beyond everything else, which is the thought that we need to love someone and are loved in return. So if the digital era is making certain kinds of human interaction inconvenient or expensive, we should start to restrain it. We should not always have to adjust to changing technology either. Technology should sometimes be forced to adapt to us. Even the environment is being roped in to help sell the argument that less human contact, courtesy of virtual meetings and relationships, is positive because it reduces our carbon footprint. This is a ridiculous suggestion: computers and virtual worlds use energy themselves (5.3 percent of global energy, to be precise), and to suggest that you can do away with physical movement and interaction altogether is crazy.

Lack of physical interaction is bad news for people and ideas too. New ideas need people to have meaningful conversations and relationships. Embryonic ideas also need people to interact with other people so that ideas can be fed, clothed, polished, and ultimately sent off into the world to fend for themselves.

It seems to me that what people seem to want more than ever these days is the opportunity to be touched emotionally by the thinking and experiences of other people. But if we all become addicted to virtual experiences and relationships, what will become of this hunger? What will happen if every individual on earth is able to personalize their experience of the world? What then of collective memory or culture, and how can humankind possibly have meaningful and cohesive discussions about important issues if we are all off doing our own thing?

Reclaiming the time and space to think

What should we, as individuals, do if we are concerned about the invasion of screen culture into our everyday lives? Bluntly, we should think.

We should think deeply about our use of digital devices and about our relationships with each other. For example, we should patronise places because of what they *don't* offer rather than because of what they do. Public libraries and schools are a case in point. They should embrace digital learning, but they should primarily celebrate their physical spaces and the people, objects, and artifacts contained within them. At least both should fight to keep some of their physical spaces free from electronic clutter and noise. And if they don't, we should make an awful lot of noise about it.

The same goes for work. We should work less. We should do things less quickly and think more slowly. We should dispense with high-performance team-building events and go out for long lunches with the people we work with instead. We should take our work brains on little adventures and set aside some mind space once in a while too. And we should, as neuroscientist Kalina Christoff suggests, make flypaper from our unfocused minds to trap new ideas and unexpected associations, because this can sometimes be more effective than methodical reasoning.

We should start thinking more seriously about the long-term benefits of doing things that appear to have no immediate value. Looking out of the window would be a good place to start. When did you last see an adult looking out of an airplane window for a long period? Come to think of it, when did you last see a child looking out of a car or train window? Almost never, I'd wager, because all eyes are usually focused downward at some kind of digital device. If looking out of windows doesn't work for you, go for a long walk. Get lost in the wilderness once in a while. Above all, spend time slowly experimenting, tinkering, analyzing, and discussing.

At home we should try to stop work from overstaying its welcome. Work will continue to intrude, but we should create a strong physical separation between clocking on and clocking off. We should also relax more. This means taking one proper day off per week (ideally two) and taking real holidays (for at least a week) at least once a year. It also means "No BlackBerry" holidays, digital diets, more time spent alone in soapy baths and garden sheds, and more sleep.

Sleep is vital not only to our physical and mental health, but because it is when we sleep that our mind starts to lay down memories and connect seemingly disparate pieces of information and experience. As the writer William Golding once said, "Sleep is where all the unsorted stuff comes flying out as from a dustbin upset in a high wind." Thus, if we are building physical environments or societal norms that do not value sleep, we are again eroding our ability to dream and think up new ideas.

So if you are convinced that deep thinking spaces are under threat, do something about it. Fight for the right not to take work home or set up a protest group to demand that airlines and restaurants have cellphone-free zones. If the paucity of dreaming time worries you, dream up ways of solving the problem.

This applies to adults but it applies to children too. Kids are overscheduled and overworked. Free play has all but disappeared. So let's invent "do nothing" days, where there is no plan and no structured play. Let's ensure that kids get bored every so often and have to use their imaginations to invent ways to escape.

Not so easy

Another potential threat to sagacity is the availability of easy pleasures. Unless we instill the sense that hard or time-consuming things can be rewarding, mindless hedonism (courtesy of the digital era) will flourish and a substantial part of the population will be stranded in a state of semi-infancy. So if something is easy, don't do it. Focus instead on things that are hard. Will this be popular? No. You could be hated for it. But over the longer term you might be thanked.

According to one survey, in China 66 percent of people believe that "it's possible to have real relationships purely online." In other words, it's possible to use technology as an alternative to direct physical interaction. But just because it's possible shouldn't mean it's preferable. Again, people are going for the easy option, but what is the end result likely to be? Our technology is becoming increasingly sophisticated but in some respects we are going backward in the quality of our communications. Convenience is becoming the measure of all things and everything is getting to be an extension of personal need and desire.

You can see a similar trend in schools. Teachers pander to young students, continually trying to engage their personal desires by offering them a smorgasbord of easy options and alternatives. But this approach simply reflects back to the students the idea of their own importance, that they are at the center of the universe, and gives weight to their personal views, no matter how ignorant or inaccurate.

Cultural critic Lee Siegel has observed: "The internet is the first environment to serve the needs of the isolated, elevated, asocial individual." Or, as English writer Tim Adams has commented, the internet "almost invariably proved itself to be an occasion for the world to understand me, rather than me understanding the world."

We think that we are using the internet, but perhaps it will end up using us. We generally assume that knowledge increases over time. The internet, we therefore assume, is spreading knowledge. But it is always possible that the reverse could be happening. Ignorance could be increasing over time because the sheer volume of digital dross and distraction that is now so easily co-created and distributed is drowning out learning and wisdom.

So perhaps we are not in the middle of an information revolution, but rather at the start of a machine-driven disinformation revolution. An electronic era in which ordinary individuals become so confused that they just give up thinking in any meaningful way. An age in which we are so focused on ourselves that our ability to relate to other people starts to decline. This could happen without us really noticing it. It will happen slowly and we will get used to it. But just because we get used to something doesn't mean that it is good.

Scientific historian George Dyson said in his book *Darwin among the Machines*: "In the game of life and evolution there are only three players at the table: human beings, nature and machines. I am firmly on the side of nature. But nature, I suspect, is on the side of the machines." It's possible, of course, that mankind has had his day and that it is now time to hand things over to the machines. But I am not going to give up on mankind (or kind man) quite so fast.

FROM SMALL BEGINNINGS

Change usually starts with individuals, so here are a few examples of individuals and groups who are fighting for a different relationship with the digital age. People who are arguing, either publicly or privately, that our sense of movement is an illusion. That our progress is circular, not linear. That we are not applying the lessons of the past nor learning from the future.

- *The Idler* is a British magazine that celebrates loafing around as an antidote to the extreme busyness of the digital era. In Japan, the Sloth Club is a similar enterprise. It encourages people to slow down and to use as little energy and other resources as possible.
- Poladroid (sic) is a computer program that allows people to replicate the look of old Polaroid photographs—or "clocks for seeing," as someone once put it. Why would anyone want to do something so anachronistic? The answer is that some people are feeling rather uneasy about the way digital technology is removing us from reality and is encouraging us to forget. Muxtape does much the same thing with MP3 files that replicate mix tapes on cassette, while Telegramstop turns immaterial (and thus forgettable) text messages into physical telegrams.
- The Clock of the Long Now Foundation in the US is dedicated to fostering long-term thinking and responsibility over a timeframe of the next 10,000 years. Central to this is the idea of a mechanical clock that, in the words of Danny Hillis, the computer scientist who dreamed up the idea, ticks once a year, bongs every century, and cuckoos every millennium. It is a meta-clock that changes our concept of time.
- Jem Finer's Long Player Project is a musical score that is intended to play, or be played, over a 1,000-year period. That's right, a piece of transgenerational art that reframes our thinking about the here and now.

Undoubtedly, with the development of robotics, artificial intelligence, genetics, and nanotechnology, machines will become more like people and people will become more like machines. But currently, only human beings can think deeply. At the moment only people have curiosity and imagination and only people build smart machines, so we are still the only ones who can invent the future. The future is full of fantastic possibilities and we can invent it any way we like if we just put our minds to it.

Are we on the cusp of a second renaissance where new ideas are widely created, discussed, and disseminated, as writers such as Clay Shirky and Nick Bilton suggest; or are we at the beginning of a new digital dark age where constant connectivity facilitates cultural conformity and leads to electronic enslavement and the death of original thought? Are we creating a world in which people find it increasingly difficult to deal with ambiguity and one where problem-solving abilities decline? I don't know the answer to such questions, but we should spend sufficient time looking for precisely the right question rather than trying to find easy answers.

Digital technology is a wonderful invention. But my intuitive feeling is that we have started to ask too much of it. If everything becomes too easy, our minds and our bodies will eventually lose their muscularity, resilience, and creativity. Hence the need to balance the fast with the slow, the analogue with the digital, the physical with the virtual, and the local with the remote. We also need to balance the ancient with the modern and learn how to use machine intelligence *in combination with* human intelligence and not as a replacement. But before we can do any of this we will need to change the way we think.

I think the jury is still very much out on whether the invasion of digital machines into our everyday lives is having serious effects on our minds. On one level we are getting smarter. Information is more widely available and we are getting clever about finding it. On the other hand, there is evidence that deep, reflective thinking is seriously threatened. There is also the very persuasive argument that a life lived digitally, at a physical distance from others, is ultimately unbearable. Nevertheless, I think more people are starting to see this. More people are realizing that our pathologically short attention spans are getting us into serious trouble and that we seem to be increasingly unable to think rigorously about the longer-term consequences of our actions.

I would like to leave you with one final thought. Technology is not destiny. The human brain is probably the most complex structure in the universe, but it has one very simple feature. It is not fixed. It is malleable. Indeed, it is impressionable to the point where it records everything that charms, deters, or touches it. You might think that certain things don't affect you, but you'd be wrong. New objects and environments are already influencing how you think, even if you don't realize it yet. However, currently we are still the smartest things around and if we do not like what we can see of the future, there is still time to change it. It might be a good idea to sit down, open a window, and ponder this thought (or even this book) for a little while.

A Taste of Future Minds: 10 Predictions

⚁ Digital storage will allow us to record our entire lives via wearable devices. Our lives will therefore become fully searchable. This will have a range of implications, ranging from "memory theft" through to issues relating to the death of forgetting.

⚁ By 2020, the majority of university textbooks will have migrated to digital formats.

⚁ Oversharing of personal information, especially real-time location, will become a major issue, linked to both personal privacy and physical security. An early example of this trend is the discussion at www.pleaserobme.com.

⚁ Machines will become aware of the emotional state of their users and will adjust themselves accordingly. Meanwhile, humans will become more machine-like in their personal relationships and thinking. Hence the development of addictive cybersex and even cybermarriage (generally man meets machine).

⚁ "Peak attention" will emerge as a concept and people will start to ration their consumption of information via digital diets or by employing professional information sifters. Similarly, information trust will become a huge issue. The semantic web will partly solve this problem, but a better solution will be to start using the local public library.

◨ A Slow Thinking movement will emerge as a parallel to the Slow Food movement, with people celebrating slow reading, slow writing, and other forms of old-fashioned paper-based communication. Some companies will also revert to paper for strategic documents due to cyberattacks and because of mistakes relating to speed-reading important information on screen.

◨ Internet addiction will become recognized as a problem, but only by governments, which will set up internet addiction clinics and cybercounseling centers. Most people, electronically tethered to the office or fanatically Facebooking friends, will refuse even to acknowledge that there is a problem. Meanwhile cyberbullying, identity theft, and medical identity theft will continue to cause big problems in the real world.

◨ Mental privacy will become a major issue, particularly for those living and working in virtual worlds.

◨ In the near future, time and space will be luxuries. We will therefore see the development of communications-free holiday resorts and quiet thinking rooms or zones in offices, libraries, hotels, airplanes, and cafés.

◨ There will be a growing convergence between digital and biological worlds, with online attitudes and behaviors influencing real-life societal norms. Direct brain-to-machine communication will drive this trend, as will other developments in haptics technology, augmented reality, and ambient intelligence.

Notes

"If your mind is empty, it is always ready for anything, it is open to everything. In the beginner's mind there are many possibilities, in the expert's mind there are few."

Shunryu Suzuki

Overture

1   **In the US, adults were spending double the amount of time online**: Forrester Research, "Consumers' behavior online: A 2009 deep dive," July 27, 2009, www.forrester.com.

1   **In Europe, the amount of time adults were online:** Forrester Research, "A deep dive into European consumers' online behaviour, 2009," August 13, 2009, www.forrester.com.

1   **Nonworking women are passing almost half their leisure time online**: *Guardian Weekend* magazine, quoting TNS study in "Of mice and women," Zoe Williams, 17 July 2010.

1   **The average person spends 45 percent of their waking hours:** Ofcom, "Communications Market Report," August 2010.

1   **The finding in 2010**: Hal Crowther, "Invasion of the mind-snatchers," *Saturday Telegraph Review*, 14 August 2010.

2   **The brain receives a blast of dopamine in the prefrontal cortex**: *Australian*, 14–15 June 2008, "Society hard-wired for a fall" by Peter Wilson.

5   **The average person's daily intake of information was 300 percent greater**: *New York Times*, 6 June 2010, "Hooked on gadgets and paying a mental price," by Matt Richtel.

7   **Online collectivism:** The Edge, 30 May 2006, "Digital Maoism." Also "The hazards of the new online collectivism" by Jaron Lanier (edge.org).

Chapter I

11   **An average of 2,272 text messages a month:** Over the Air blog, *Information Week*, May 2009.

11   **A 2010 report:** Ofcom, "Communications Market Report," August 2010.

11   **Kaiser report:** *Scientific American Mind*, October/November 2008, "Meet your iBrain" by Gary Small & Gigi Vorgan.

11   **Consumption of printed media is generally in decline:** Bureau of Labor Statistics, "American Time Use Survey," www.bls.gov/tus/.

13   **A study by the Social Issues Research Center:** *The Australian*, 31 December 2005, "The I-want-it-now years," D. Hope, www.theaustralian.com.au; *Sydney Morning Herald*, 17–18 December 2005, "The modern dream... life is meant to be easy," www.smh.com.au.

14   **Ambient intimacy:** *New York Times*, 7 September 2008, "Brave new world of digital intimacy" by Clive Thompson.

16   **Micro boredom:** *Sydney Morning Herald*, 21 March 2008, "My iPod ate my brain" by Carolyn Johnson.

16   **A couple in South Korea:** *Daily Telegraph*, 6 March 2010, "Internet obsessed pair 'let baby die.'"

16   **Internet addiction:** *The Author*, Spring 2009, "Google Me Stupid" by Rita Carter.

19   **Culturally induced schizophrenia:** Fredric Jameson (1991) *Postmodernism, or The Cultural Logic of Late Capitalism*, Durham: Duke University Press.

19 **To become what Hal Crowther terms "blessedly discon-nected":** "Invasion of the mind-snatchers," *Saturday Telegraph Review*, 14 August 2010.

20 **An Ofcom report in 2010:** Ofcom, "Communications Market Report," August 2010.

20 **Multitasking increases stress-related hormones**: *The Atlantic*, November 2007, "Autumn of the multi-taskers" by Walter Kirn.

21 **People who are continually distracted by email suffer from higher levels of stress**: *New York Times*, 6 June 2010, "Hooked on gadgets and paying a mental price," by Matt Richtel.

21 **Such stress can be linked to lower levels of short-term memory**: *Ibid.*

21 **To quote Prensky**: Mark Bauerlein (2008) *The Dumbest Generation*, New York: Jeremy P Tarcher.

22 **Naomi Baron claims there is an "intellectual torpor"**: *The Author*, Spring 2009, "Google me stupid" by Rita Carter.

22 **Chaucer couldn't spell in today's terms**: *Economist*, 10 April 2008, "Homo Mobilis."

23 **Socrates**: *Sydney Morning Herald*, 3–4 January 2009, "The net effect" by Richard King.

24 **Tree octopus**: *International Herald Tribune*, "Literacy debate: Online, r u really reading?" by Motoko Rich.

24 **Jacob Nielsen study**: *Chronicle of Higher Education*, 29 September 2008, "Online literacy is a lesser kind" by Mark Bauerlein.

25 **90 percent of employers think "reading comprehension" is "very important"**: *International Herald Tribune*, "Literacy debate: Online, r u really reading?" by Motoko Rich.

25 **IQ scores rising; Flynn effect**: *Wired*, 13 May 2005, "Dome improvement" by Steven Johnson.

26 **Diet affecting intelligence**: *Scientific American Mind*, Feb/March 2009, "Six ways to boost brainpower."

26 **Millions of workers in the UK and the US are functionally illiterate**: *Weekly Telegraph*, 5–11 February 2009, "Half of us are barely literate and it's getting worse" by Alistair Palmer.

Chapter 2

28 **2,000 hours in front of a screen**: Susan Greenfield talk at the Centre for Independent Studies, September 2009, quoted in *Sydney Morning Herald*, 12–13 September 2009, "We're losing our minds over technology" by Miranda Devine.

29 **The Net may well be the single most powerful mind-altering technology**: *Nicholas Carr (2010) The Shallows: What the Internet Is Doing to Our Brains*, New York: WW Norton, pp. 35, 116.

29 **Prescriptions for Ritalin to treat hyperactivity**: *Australian*, 14–15 June 2008, "Society hard-wired for a fall" by Peter Wilson.

32 **49 percent of British children banned by their parents from climbing trees**: *The Week*, 9 August 2008.

32 **Surgeons who play video games**: *Scientific American Mind*, Feb/March 2009, "Six ways to boost brainpower."

33 **The average amount of time a father spends alone with his child**: JP Robinson, VG Andreyenkov, & VD Patrushev (1988) *The Rhythm of Everyday Life*, Boulder, CO: Westview Press.

34 **Older children consistently claim that they feel hurt**: *Observer*, 27 June 2010, quoting Dr. Sherry Turkle in the *New York Times*.

35 **No correlation between increased expenditure on technology and improved student learning:** *Chronicle of Higher Education*, 29 September 2008, "Online literacy is a lesser kind" by Mark Bauerlein.

35 **Scottish report; Munich University report:** both quoted in Bauerlein, *Dumbest Generation*.

35 **One 2010 report co-authored by Professor Jacob Vigdor at Duke University:** *Daily Telegraph*, 22 June 2010, "Love of reading at risk, says Tom Stoppard."

36 **Why books still matter:** *The Atlantic*, 2 March 2009, "Resisting the Kindle" by Sven Birkerts; *The Times*, 7 May 2009, "Pulp friction: The crisis facing the book" by Nicholas Clee; *Chronicle of Higher Education*, 19 Sept. 2008, "Online literacy is a lesser kind" by Mark Bauerlein; *Australian Literary Review*, 1 July 2009, "Is that a canon in your pocket?" by Geordie Williamson.

37 **A Kaiser report in the US called "Generation M":** quoted in Bauerlein, *Dumbest Generation*.

37 **25 percent of work handed in by pupils contains material copied directly from the internet:** BBC News, 6 November 2008, "Teachers voice plagiarism fears."

38 **Multitasking is becoming an increasingly normal state:** National Public Radio Science Friday interview by Paul Raeburn, "Multi-tasking may not mean higher productivity," 7 July 2010.

38 **Open-source textbooks:** *Scientific American*, 14 August 2009, "Open-source textbooks a mixed bag in California" by Brendan Borrell.

38 **Lack of resilience:** Robert Hughes (1994) *Culture of Complaint*, New York: Grand Central Publishing.

Chapter 3

49   **Roughly what you'd find inside a $5 calculator**: *Guardian*, 12 March 2005, "Messiness of the mind" by Steven Rose.

49   **Differences between mind and machine**: Spiked Online, 3 July 2008, "The World Wide Web is nothing like a brain" by Stuart Derbyshire.

52   **Like flying over Los Angeles at night**: *The Atlantic*, July/August 2008, "My amygdala, my self" by Jeffrey Goldberg.

55   **Ritalin and Provigil**: *Economist*, 24 May 2008, "Cognitive enhancement: All in the mind."

59   **An organization called Cephos**: *The Times* magazine, 28 February 2009, "What if a stranger could read your mind?" by John Naish.

Chapter 4

64   **Neuroscience of leadership**: *Strategy + Business*, "The Neuroscience of Leadership" by David Rock and Jeffrey Schwartz.

65   **The brain is more receptive to new information when we are in a good mood**: *Scientific American*, 18 December 2006, "Happiness: Good for creativity, bad for single-minded focus" by JR Minkel.

66   **Early blockages to thinking**: *Scientific American*, 25 January 2008, "What are we thinking when we (try to) solve problems?' by Nikhil Swaminathan.

68   **Arthur Koestler summed this up when he said**: *Prospect*, October 2008, "Speculations" by Jim Holt.

69   **At age 6 our brain is already 95 percent of its adult weight**: *New Scientist*, 4 April 2009, "The five ages of the brain."

69 **A slow but terminal decline**: *Ibid.*

69 **Habituation***: Ibid.*

72 **Split-brain theory**: *Scientific American Mind*, June/July 2008, "Spheres of influence" by Michael Gazzaniga.

73 **It might be better to think of two brains**: *Weekly Telegraph*, 17–23 February 2010, "Adults have edge over internet generation" by Martin Evans.

73 **Male and female brains**: *Scientific American*, 26 January 1998, "Is it true that creativity resides in the right hemisphere of the brain?"

75 **Sometimes misleading memories attach themselves to information**: *Harvard Business Review*, February 2009, "Why good leaders make bad decisions" by Andrew Campbell, Jo Whitehead, & Sydney Finkelstein.

79 **Lack of exercise can inhibit the formation of new brain cells**: *Scientific American Mind*, Feb/March 2009, "Six ways to boost brainpower."

79 **Do you get enough sleep**: *Scientific American*, 7 August 2008, "Sleep on it: How snoozing makes you smarter" by Robert Stickgold & Jeffrey Ellenbogen; *Guardian*, 30 December 2006, "How did you sleep last night?" by John Crace.

80 **Daydreaming links to meditation**: *Scientific American Mind*, Feb/March 2009, "Six ways to boost brainpower."

82 **James Surowiecki:** *Economist*, 28 June 2008, "The crowd within."

83 **Physical teams of two**: *Harvard Business Review*, May 2009, "Why teams don't work" by Diane Coutu, www.hbr.org.

Chapter 5

88 **Quote from Ed Mitchell**: *The Times*, 20 July 2009, "One small step back to where it started" by Mark Mason.

101 **Piers are about dreaming and reflection**: *Economist*, 22 September 2007, "The end of the pier."

101 **Gardens**: *Sydney Morning Herald*, 24–25 March 2007, "I garden, therefore I am" by Cheryl Maddocks.

104 **Germaine Greer**: *Guardian*, 28 October 2006, "It's about death as much as it is about life."

104 **Private gardens are not generally expansive nowadays**: *Sydney Morning Herald*, 19–20 August 2006, "Goodbye to the great outdoors" by Nick Galvin.

104 **A study by Marc Berman**: *Sun Herald*, 1 March 2009, "It's a jungle out there" by Simon Webster.

105 **Open-plan office research**: *Sydney Morning Herald*, 30 April 2009, "Office warfare" by Terry Smyth; research by Dr. Vinesh Oommen at the Queensland University of Technology; *Asia Pacific Journal of Health Management*, January 2009.

105 **60-foot radius rule**: *New Yorker*, 11 December 2000, "Designs for working" by Malcolm Gladwell.

106 **Design of Pixar headquarters**: *Harvard Business Review*, September 2008, "How Pixar fosters collective creativity" by Ed Catmull.

108 **Saul Greenberg**: *Sydney Morning Herald*, May 31–June 1, 2008, "A space of your own" by Owen Thomson.

109 **National Association of Professional Organizers**: *Economist*, 21 June 2008, "DNA all over the place."

110 **22 percent of employees used their iPods at work**: *Sydney Morning Herald*, 26–28 January 2007, "An Apple for the teacher" by Lisa Timson.

111 **Soundscapes**: *Financial Times*, "Sound discovery" by Harry Eyres.

112 **BOSTI research**: *Business 2.0*, April 2007, "Thinking outside the cube" by Jeffrey Pfeffer.

113 **Frank Gehry was hired to build the new Beckman Center**: *Wired*, December 2006.

114 **"active" or "inactive" fruit**: *Weekly Telegraph*, 10–16 January 2007, "Taxman 'Shells out £7m on tidy desks project'" by Paul Stokes.

115 **A little disorder can be healthy**: *Economist*, 19 December 2002, "In praise of clutter."

115 **Leon Heppel**: *Financial Times*, 11 January 2006, "A certain level of chaos? Well worth organising..." by David Honigmann.

119 **Best Buy**: *Business Week*, 11 December 2006, "Smashing the clock" by Michelle Conlin.

124 **Porches**: *Financial Times*, 27–28 October 2007, "The threshold of greatness" by Phillip Gulley.

127 **Bedouin workers**: *Economist*, 12 April 2008, "Our nomadic future."

128 **The list of major inventions created by accident**: *Wired*, April 2008, "Unlikely places where Wired pioneers had their eureka! moments" by Mathew Honan.

134 **We need to develop unhurried minds**: News.com.au, 3 June 2009, "Information overload from Twitter, Facebook, TV robs us of compassion—scientists."

Chapter 6

149 **Constant partial attention**: *New York Times*, 16 October 2005, "Meet the life hackers" by Clive Thompson.

149 **Chalupa's radical idea**: World Question Centre (edge.org), Leo Chalupa.

150 **Consider what Bill Gates used to do**: *Wall Street Journal*, 28 March 2005, "In a secret hideaway, Bill Gates ponders Microsoft's future" by Robert Guth.

152 **Latent inhibition**: Cindy Rabe (2006) *The Innovation Killer*, New York: Amacom, p. 81.

153 **Losing your inhibitions**: *Scientific American*, 24 March 2005, "Unleashing creativity" by Ulrich Kraft.

162 **Internet uses 5.3 percent of global energy**: Kevin Kelly, TED talk, 17 October 2008.

164 **Make flypaper from our unfocused minds**: Fast Company.com, "Hard work's overrated, maybe detrimental" by Cliff Kuang.

165 **66 percent of people believe it's possible to have real relationships purely online**: *Economist*, 7 February 2009, "Virtual pleasures."

166 **An occasion for the world to understand me**: *Observer*, 6 December 2009, "Can the art of great writing survive the digital age?"

166 **Scientific historian George Dyson said**: *Sunday Life*, 6 April 2009, "Switching Off" by Janelle McCulloch.

169 **The human brain is probably the most complex structure in the universe**: *Economist*, "A survey of the brain," 23 December 2006.

Bibliography

"The meaning of things lies not in the things themselves, but in our attitude towards them."

Antoine de Saint-Exupery

Abrahamson, Eric & Hallowell, Edward (2006) *A Perfect Mess: The Hidden Benefits of Disorder*, London: Weidenfeld & Nicolson.

Allen, David (2001) *Getting Things Done: The Art of Stress-Free Productivity*, Harmondsworth: Penguin.

Ariely, Dan (2010) *Predictably Irrational: The Hidden Forces that Shape Our Decisions*, New York: HarperCollins.

Ball, Philip (2010) *The Music Instinct: How Music Works and Why We Can't Do Without It*, Oxford: Bodley Head.

Baron, Naomi (2008) *Always On: Language in an Online and Mobile World*, New York: Oxford University Press.

Bauerlein, Mark (2009) *The Dumbest Generation: How the Digital Age Stupefies Young Americans and Jeopardizes Our Future (Or, Don't Trust Anyone Under 30)*, Los Angeles, CA: Jeremy P Tarcher.

Bennis, Warren & Biederman, Patricia Ward (1998) *Organizing Genius: The Secrets of Creative Collaboration*, London: Nicholas Brealey Publishing.

Bilton, Nick (2010) *I Live in the Future and Here's How It Works: Why Your World, Work, and Brain Are Being Creatively Disrupted*, New York: Crown Business.

Blyth, Catherine (2009) *The Art of Conversation: Or, What to Say and When*, London: John Murray.

Boden, Margaret (2003) *The Creative Mind: Myths and Mechanisms*, London: Routledge.

Brand, Stewart (2000) *The Clock of the Long Now: Time and Responsibility*, New York: Basic Books.

Brand, Stewart (1997) *How Buildings Learn: What Happens After They're Built*, London: Weidenfeld & Nicolson.

Brin, David (1999) *The Transparent Society: Will Technology Force Us to Choose Between Privacy and Freedom?* New York: Basic Books.

Brockman, John (ed.) (2007) *What Is Your Dangerous Idea? Today's Leading Thinkers on the Unthinkable*, New York: Pocket Books.

Broderick, Damien (1999) *The Last Mortal Generation*, Reed Natural History Australia.

Brown, Shona L & Eisenhardt, Kathleen M (1998) *Competing on the Edge: Strategy as Structured Chaos*, Boston, MA: Harvard Business School Press.

Buzan, Tony (1995) *The Mindmap Book*, London: BBC Books.

Carr, Nicholas (2010) *The Shallows: What the Internet Is Doing to Our Brains*, New York: WH Norton.

Carter, Rita (2010) *Mapping the Mind*, London: Phoenix.

Carter, Rita (2010) *Exploring Consciousness*, Berkeley, CA: University of California Press.

Castiglione, Baldassare (2002) *The Book of the Courtier*, London: Norton Critical Editions.

Christensen, Clayton (2003) *The Innovator's Dilemma: The Revolutionary Book that Will Change the Way You Do Business*, London: HarperBusiness Essentials.

Churchland, Paul (1996) *The Engine of Reason, The Seat of the Soul: Philosophical Journey into the Brain*, Boston, MA: MIT Press.

Clippinger, John Henry (2007) *A Crowd of One: The Future of Individual Identity*, London: Public Affairs.

Clotfelter, Charles T, Ladd, Helen F, & Vigdor, Jacob L (2009) "Scaling the digital divide: Home computer technology and student achievement," Durham, NC: Duke University.

Colvin, Geoff (2008) *Talent is Overrated: What Really Separates World-Class Performers from Everybody Else*, London: Nicholas Brealey Publishing.

Crowther, Hal (2010) One Hundred Fears of Solitude, *Granta*, Issue 111.

Crystal, David (2008) *Txtng: The Gr8 Db8*, Oxford: Oxford University Press.

Davies, Mike (2007) *The Monster at Our Door: The Global Threat of Avian Flu*, New York: The New Press.

Davies, Nick (2009) *Flat Earth News: An Award-winning Reporter Exposes Falsehood, Distortion and Propaganda in the Global Media*, London: Vintage.

Davis-Floyd, Robbie & Arvidson, P Sven (eds) (1997) *Intuition: The Inside Story*, London: Routledge.

Dawson, Ross (2008) *Living Networks: Leading Your Company, Customers, and Partners in the Hyper-Connected Economy*, Lulu.com.

de Bono, Edward (2009) *Six Thinking Hats*, Harmondsworth: Penguin.

de Bono, Edward (1990) *The Use of Lateral Thinking*, Harmondsworth: Penguin.

Dennett, Daniel (1996) *Darwin's Dangerous Idea: Evolution and the Meanings of Life*, Harmondsworth: Penguin.

Doidge, Norman (2008) *The Brain that Changes Itself: Stories of Personal Triumph from the Frontiers of Brain Science*, Harmondsworth: Penguin.

Dru, Jean-Marie (1996) *Disruption: Overturning Conventions and Shaking Up the Marketplace*, Chichester: John Wiley.

Dyson, George (1999) *Darwin Among the Machines*, Harmondsworth: Penguin.

Fernandez-Armesto, Felipe (2005) *Ideas that Changed the World*, London: Dorling Kindersley.

Ferriss, Timothy (2007) *The 4-Hour Workweek: Escape the 9–5, Live Anywhere and Join the New Rich*, New York: Vermilion.

Fine, Cordelia (2007) *A Mind of Its Own: How Your Brain Distorts and Deceives*, New York: Icon Books.

Florida, Richard (2003) *The Rise of the Creative Class: And How It's Transforming Work, Leisure, Community and Everyday Life*, New York: Basic Books.

Gare, Shelley (2005) *The Triumph of the Airheads*, South Paris, ME: Park Street Press.

Gershenfeld, Neil (2000) *When Things Start to Think*, New York: Owl Books.

Gigerenzer, Gerd (2008) *Gut Feelings: Short Cuts to Better Decision Making*, Harmondsworth: Penguin.

Gladwell, Malcolm (2006) *Blink: The Power of Thinking without Thinking*, Harmondsworth: Penguin.

Gleick, James (2000) *Faster: The Acceleration of Just about Everything*, New York: Vintage Books.

Goleman, Daniel (1996) *Emotional Intelligence: Why It Can Matter more than IQ*, London: Bloomsbury.

Gosden, Roger (2000) *Designer Babies: Science and the Future of Human Reproduction*, London: Phoenix.

Greenfield, Susan (2004) *Tomorrow's People: How 21st-Century Is Changing the Way We Think*, Harmondsworth: Penguin.

Greenfield, Susan (2009) *ID: The Quest for Meaning in the 21st Century*, London: Sceptre.

Grudin, Robert (1991) *The Grace of Great Things: On the Nature of Creativity*, Orlando, FL: Houghton Mifflin.

Hallinan, Joseph T (2010) *Why We Make Mistakes: How We Look Without Seeing, Forget Things in Seconds, and Are All Pretty Sure We Are Way Above Average*, New York: Broadway Books.

Hallowell, Edward (2007) *Crazy Busy: Overstretched, Overbooked, and About to Snap! Strategies for Handling Your Fast-Paced Life*, New York: Ballantine Books.

Hargadon, Andrew (2003) *How Breakthroughs Happen: The Surprising Truth about How Companies Innovate*, Boston, MA: Harvard Business School Press.

Harkin, James (2009) *Cyburbia: The Dangerous Idea that's Changing How We Live and Who We Are*, Boston, MA: Little, Brown.

Harper, Richard and Sellen, Abigail (2003) *The Myth of the Paperless Office*, Boston, MA: MIT Press.

Hodgkinson, Tom (2007) *How to Be Free*, Harmondsworth: Penguin.

Honore, Carl (2005) *In Praise of Slow: How a Worldwide Movement is Challenging the Cult of Speed*, London: Orion.

Iggulden, Conn & Iggulden, Hal (2006) *The Dangerous Book for Boys*, London: HarperCollins.

Jackson, Maggie (2010) *Distracted: The Erosion of Attention and the Coming Dark Age*, Tonbridge: Prometheus.

Johnson, Steven (2006) *Everything Bad Is Good for You: How Popular Culture Is Making Us Smarter*, Harmondsworth: Penguin.

Jones, Benjamin F. (2008) "The burden of knowledge and the death of the renaissance man: Is innovation getting harder?" *Review of Economic Studies*, 76(1): 283–317.

Joy, Bill (2000) "Why the future doesn't need us," *Wired*, 8.04 (April), www.wired.com/wired/archive/8.04/joy.html.

Kao, John (1997) *Jamming: The Art and Discipline of Business Creativity*, London: HarperCollins.

Keen, Andrew (2008) *The Cult of the Amateur: How Blogs, MySpace, YouTube and the Rest of Today's User-Generated Media Are Killing Our Culture and Economy*, London: Nicholas Brealey Publishing.

Kim, Chan & Mauborgne, Renee (2005) *Blue Ocean Strategy: How to Create Uncontested Market Space and Make the Competition Irrelevant*, Boston, MA: Harvard Business School Press.

Koestler, Arthur (1975) *The Act of Creation*, London: Picador.

Kuhn, Thomas (1996) *The Structure of Scientific Revolutions*, Chicago, IL: Chicago University Press.

Kurzweil, Raymond (2006) *The Singularity Is Near: When Humans Transcend Biology*, London: Duckworth.

Kusek, David & Leonhard, Gerd (2005) *The Future of Music: Manifesto for the Digital Music Revolution*, London: Omnibus.

Lammiman, Jean & Syrett, Michel (1998) *Innovation at the Top: Where Do Directors Get Their Ideas?* Horsham: Roffey Park Institute.

Lawrence, TE (1997) *Seven Pillars of Wisdom*, Oxford: Wordsworth Editions.

Lehrer, Jonah (2010) *The Decisive Moment: How the Brain Makes Up its Mind*, London: Canongate Books.

Leonard, Dorothy & Swap, Walter (2005) *When Sparks Fly*, Boston, MA: Harvard Business School Press.

Levitin, Daniel (2008) *This Is Your Brain on Music: Understanding a Human Obsession*, London: Atlantic Books.

Levitin, Daniel (2010) *The World in Six Songs: How the Musical Brain Created Human Nature*, London: Aurum Press.

Lienhard, John (2008) *How Invention Begins: Echoes of Old Voices in the Rise of New Machines*, Oxford: Oxford University Press.

Lynch, Gary & Granger, Richard (2009) *Big Brain: The Origins and Future of Human Intelligence*, London: Palgrave Macmillan.

Mackay, Charles (1995) *Extraordinary Popular Delusions and The Madness of Crowds*, London: Wordsworth Editions.

Maitland, Sara (2009) *A Book of Silence*, London: Granta Books.

Marcus, Gary (2008) *Kluge: The Haphazard Construction of the Human Mind*, London: Faber & Faber.

Markova, Dawna (1997) *The Open Mind: Exploring the Six Patterns of Natural Intelligence*, Newburyport, MA: Conari Press.

Merriam, Alan (1964) *The Anthropology of Music*, Evanston, IL: Northwestern University Press.

Merrill, Douglas & Martin, James (2010) *Getting Organized in the Google Era: How to Get Stuff Out of Your Head, Find It When You Need It, and Get It Done Right*, New York: Broadway Books.

Miller, Arthur (2000) *Insights of Genius: Imagery and Creativity in Science and Art*, Boston: MIT Press.

Miller, Arthur (2002) *Einstein, Picasso: Space, Time and the Beauty that Causes Havoc*, New York: Basic Books.

Morgan Stanley Research (2009) "How teenagers consume media," 10 July.

Naish, John (2009) *Enough: Breaking Free from the World of More*, London: Hodder.

Ornstein, Robert (1986) *The Psychology of Consciousness*, Harmondsworth: Penguin.

Ornstein, Robert (1997) *The Right Mind: Making Sense of the Hemispheres*, Orlando, FL: Houghton Mifflin Harcourt.

Palmer, Sue (2008) *Detoxing Childhood: What Parents Need to Know to Raise Bright, Balanced Children*, London: Orion.

Pélegrin-Genel, Elisabeth (1996) *The Office*, Paris: Flammarion.

Pink, Daniel (2008) *A Whole New Mind: Why Right-Brainers Will Rule the Future*, Oxford: Marshall Cavendish.

Pollan, Michael (2008) *A Place of My Own: The Architecture of Daydreams*, Harmondsworth: Penguin.

Powers, William (2006) "Hamlet's Blackberry: Why Paper is Eternal," Joan Shorenstein Center, Boston, MA: Harvard University, www.scribd.com/doc/3562724/Hamlets-Blackberry-Why-Paper-Is-Eternal.

Rabe, Cynthia Barton (2006) *The Innovation Killer: How What We Know Limits What We Can Imagine—And What Smart Companies Are Doing About It*, New York: Amacom.

Reichold, Klaus & Graf, Bernhard (2004) *Buildings that Changed the World*, New York: Prestel.

Richardson, Ken (2000) *The Making of Intelligence*, London: Phoenix.

Robinson, Alan & Stern, Sam (1998) *Corporate Creativity: How Innovation and Improvement Actually Happen*, San Francisco, CA: Berrett-Koehler.

Robinson, Ken (2010) *The Element: How Finding Your Passion Changes Everything*, Harmondsworth: Penguin.

Rock, David & Schwartz, Jeffrey (2006) "The neuroscience of leadership," *Strategy + Business*.

Rosenberg, Howard & Feldman, Charles (2010) *No Time To Think: The Menace of Media Speed and the 24-Hour News Cycle*, London: Continuum.

Sacks, Oliver (2008) *Musicophilia: Tales of Music and the Brain*, London: Picador.

Sacks, Oliver (2009) *The Man Who Mistook His Wife for a Hat*, London: Picador.

Schafer, Murray (1970) *The Book of Noise*, Petone: Price Milburn.

Schwartz, Barry (2005) *The Paradox of Choice: Why More Is Less*, London: HarperCollins.

Seidensticker, Bob (2006) *Future Hype: The Myths of Technology Change*, San Francisco, CA: Berrett-Koehler.

Shirky, Clay (2009) *Here Comes Everybody: How Change Happens When People Come Together*, Harmondsworth: Penguin.

Shirky, Clay (2010) *Cognitive Surplus: Creativity and Generosity in a Connected Age*, London: Allen Lane.

Siegel, Lee (2008) *Against the Machine: Being Human in the Age of the Electronic Mob*, London: Serpent's Tail.

Small, Gary & Vorgan, Gigi (2009) *iBrain: Surviving the Technological Alteration of the Modern Mind*, London: Harper Paperbacks.

Stafford, Tom and Webb, Matt (2004) *Mind Hacks: Tips and Tricks for Using Your Brain*, Sebastopol, CA: O'Reilly Media.

Stanovich, Keith (2010) *What Intelligence Tests Miss: The Psychology of Rational Thought*, New Haven, CT: Yale University Press.

Stein, Morris (1975) *Stimulating Creativity*, New York: Academic Press.

Steinbeck, John (2000) *Cannery Row*, Harmondsworth: Penguin Modern Classics.

Surowiecki, James (2005) *The Wisdom of Crowds: Why the Many Are Smarter than the Few*, London: Abacus.

Tallis, Raymond (2009) *The Kingdom of Infinite Space: A Fantastic Journey Around Your Head*, New York: Atlantic Books.

Tammet, Daniel (2009) *Embracing the Wide Sky: A Tour Across the Horizons of the Mind*, London: Hodder Paperbacks.

Tapscott, Don (2008) *Grown Up Digital: How the Net Generation is Changing Your World*, London: McGraw-Hill Professional.

Tenner, Edward (1997) *Why Things Bite Back: Technology and the Revenge of Unintended Consequences*, London: Vintage Books.

Thoreau, Henry David (2008) *Walden: Or, Life in the Woods*, Oxford: Oxford Paperbacks.

Turkle, Sherry (2007) *Evocative Objects: Things We Think With*, Boston, MA: MIT Press.

Turkle, Sherry (2008) *The Inner History of Devices*, Boston, MA: MIT Press.

Vedantam, Shankar (2010) *The Hidden Brain: How Our Unconscious Minds Elect Presidents, Control Markets, Wage Wars, and Save Our Lives,* New York: Spiegel & Grau.

Watson, Peter (2001) *A Terrible Beauty: A History of the People and Ideas that Shaped the Modern Mind,* London: Phoenix.

Watson, Peter (2006) *Ideas: a History from Fire to Freud,* London: Phoenix.

Weeks, David & James, Jamie (1996) *Eccentrics: A Study of Sanity and Strangeness,* New York: Kodansha.

Wolpert, Lewis & Richards, Alison (1997) *Passionate Minds: The Inner World of Scientists,* Oxford: Oxford University Press.

Zittrain, Jonathan (2009) *The Future of the Internet and How To Stop It,* Harmondsworth: Penguin.

A few interesting websites

Institute of Ideas—www.instituteofideas.com

Long Bets Foundation—www.longbets.org

Long Now Foundation—www.longnow.org

Longplayer Project—www.longplayer.org

Muxtape—www.muxtape.com

Portrait photography—www.lizhandy.net

Sloth Club—www.sloth.gr.jp/E-chapter/english1.htm

Slow Movement—www.slowmovement.com

TED—www.ted.com

Telegramstop—www.telegramstop.com

The Idler magazine—www.idler.co.uk

World Question Center—www.edge.org/questioncenter.html

The soundtrack to the book

"I can't think in sharps, I think in flats."

Elton John

Given the references to music and auditory environments in this book, I thought a few people might be interested in what was playing when I wasn't engrossed in total silence.

AC/DC, *Back in Black*
Angels, *Tour EP 2008*
Aria, *Soundtrack*
James Blunt, *All the Lost Souls*
David Bowie, *The Rise and Fall of Ziggy Stardust and the Spiders from Mars*
Maria Callas, *Puccini and Bellini Arias*
Leonard Cohen, *Live in London*
Dire Straits, *Brothers in Arms*
Bob Dylan, *Blood on the Tracks*
Everything But the Girl, *Idlewild*
Fleetwood Mac, *Rumours*
Ben Folds, *Ben Folds Live*
Jan Garbeck, *Officium*
Marvin Gaye, *What's Going On*
Genesis, *And Then There Were Three*
Howard Goodall, *Choral Works*
Henryk Gorecki, *Symphony No. 3*
Geoffrey Gurrumul Yunupingu, *Gurrumul*
The Killers, *Day and Age*
Bob Marley and the Wailers, *Babylon by Bus*
Meatloaf, *Bat Out of Hell*
George Michael, *Songs from the Last Century*

Joni Mitchell, *Blue*
Ennio Morricone, *The Mission Soundtrack*
Van Morrison, *Astral Weeks*
Michael Nyman, *Wonderland Soundtrack*
Pink Floyd, *Dark Side of the Moon*
Pink Floyd, *Wish You Were Here*
Giacomo Puccini, *Madame Butterfly*
Giacomo Puccini, *Turandot*
Otis Redding, *Otis Blue*
Lou Reed, *Transformer*
REM, *Automatic for the People*
Bic Runga, *Beautiful Collision*
Bob Seger & The Silver Bullet Band, *Greatest Hits*
Silence (classical compilation)
Bruce Springsteen, *Born to Run*
Talking Heads, *Stop Making Sense*
The Band, *The Last Waltz*
The Clash, *London Calling*
Amy Winehouse, *Back to Black*
George Winston, *Autumn*
George Winston, *Winter*

Acknowledgments

"A mind that is fast is sick. A mind that is slow is sound. A mind that is still is divine."

Meher Baba

First and foremost, I would like to say thank you to everyone at Nicholas Brealey for helping to make this book a reality. Secondly I'd like to say a very big thank-you to Sally and Corrina for keeping the manuscripts straight and true. Thirdly, I'd like to say thanks to Georgie for putting up with me. I'd also like to say thanks to Ross Dawson from whose book (*Living Networks*) I borrowed the title of Chapter 4. Also thank you to Tom Brigstocke, Wayde Bull, and Sandy Belford for the thinking space. Also, a big thank-you to Douglas Slater, Ken McBryde, Oliver Freeman, and Andrew Crosthwaite for some intelligent suggestions on various chapters and to my two sons, Nick and Matt, for forcing me to explain things to them until they made sense to me. Finally, a big thank-you to everyone else who helped with this book, especially the people who answered my questions.

Index

Also by Richard Watson

FUTURE FILES
A Brief History of the Next 50 Years

"Provocative, entertaining, full of surprising facts—a book to help you decide whether the world is going mad or possibly becoming more intelligent."
Theodore Zeldin

William Gibson meets Alvin Toffler in this lively and witty look at our possible futures. Filled with provocative forecasts about how the world might change in the next half century, Future Files examines emerging patterns and developments in society, technology, economy, and business, and makes educated speculations as to where they might take us.

Discover the most significant drivers of change—and the five things that won't change in the future. There's even an extinction timeline, which plots the demise of everything from getting lost to a good night's sleep, from glaciers to Google.

Indispensable to business analysts, strategists and organizations who need to stay ahead of the game as well as providing rich and fascinating material for dinner party conversations, its goal is to liberate our collective and individual imaginations so that we can see the familiar in a new light and the unfamiliar with greater clarity.

£9.99 UK PB 288pp ISBN 978-1-85788-534-7
www.nicholasbrealey.com